The Golden Relationship:
Art, Math & Nature

book two

THE SURFACE
PLANE

by
Martha Boles & Rochelle Newman

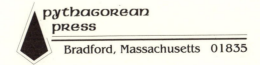
pythagorean
press
Bradford, Massachusetts 01835

THE SURFACE PLANE

All hand illustrations by Susan Libby, Newburyport, Massachusetts.

All photographs, unless otherwise indicated, are the work of Richard Newman.

ISBN 0-9614504-2-8

This book was illustrated and designed on a MacIntosh SE computer and proofed on a Laserwriter IINT printer. It was manufactured by Heffernan Press, Inc., of Worcester, Massachusetts, and printed on recycled paper, by Cross Pointe Paper Corporation.

Library of Congress Catalog Card Number: 85-60105

Cover photo: Richard Newman
New England Autumn

About the authors:
Martha Boles is Professor of Mathematics at Bradford College, Bradford, Massachusetts.
Rochelle Newman is an artist and Professor of Art at Northern Essex Community College, Haverhill, Massachusetts.
The text is one aspect of an ongoing collaboration dedicated to the concept of interdisciplinary learning.

This book is dedicated to the sense of play in which the questions "What if?" and "How come?" are central to the game.

Acknowledgements

For the works that have guided us in our quest for understanding, we are indebted to these people:

Fritjof Capra, Keith Critchlow, Gyorgy Doczi, Andreas Feininger, George Fleck and Marjorie Seneschal, Branco Grunbaum and G.C. Shephard, Robert Lawlor, Arthur Loeb, Benoit Mandelbrot, Peter Stevens, D'Arcy Thompson, David Wade.

In addition, we are grateful to:

Donna Fowler, our right arm and third head,

our assistants, without whose help and computer expertise the play might have become drudgery:
Cynthia Hastings, for her caring, perserverance and the multitude of wonderful diagrams she generated, and
Linda Maddox, who told us it couldn't be math because she was having so much fun,

Richard Newman, for the use of his art and photographic work and his all around support,

Susan Libby, for her hand drawn illustrations,

Jon Choate, for helping us with the chaos in our fractals,

Ann MacLean, for her help in proof reading, and

Sylvia Burnside for her contribution of computer generated diagrams.

Contents

Constructions

Constructions

Preface

Everything in nature is designed for a purpose, whether it is the shape and color of a flower, the placement of leaves along a branch, or the scales on the butterfly's wing...The more closely we look at objects of nature, the more beauty we find within that purposefulness of design...The beauty that is clarity of organization, economy of material, symmetry of shape, perfection of execution—qualities inherent in every form of nature.

Andreas Feininger
The Anatomy of Nature

The essential purpose of Nature is the survival of the entire system of connectedness. Each element plays a particular role in the continuance and none dominates the others. If this be so, Nature should be the model for human invention. But how can it? The majority of us live in a highly urbanized environment in which most of what we know, use, hear, smell, taste, and touch is removed from direct contact with the natural world. The foods we eat are found in jars, tins, and paper packages. The clothes we wear are heat-sealed, and permanent-pressed. The transportation we ride in is high-speed, metallic, and fiberglass. We keep ourselves warm or cool by pushing a button up or down. We learn about the weather and the state of our neighbors by looking at moving images on a screen. We eschew solitude and would rather hear electronic pulses plugged into our ears. Night becomes a sky filled with neon, while day becomes a work place inside steel and concrete. So, how can we look to Nature for our model when we hot-top everything including our feelings?

Yet Nature is there. And for us, the authors, the search for relationships and connectedness continues. Our concern is for a balance between the natural and the artificial. Our belief remains firm that Nature is the essential ground on which to focus the figures of Mathematics and Art.

If Nature does everything with a purpose and humans are part of this, what then is the purpose of humankind? If we cannot remove ourselves from the physicality of the universe, is then our purpose to reflect through our own works those qualities found in the rest of the natural world? Should not the artist and the mathematician also seek "clarity of organization, economy of materials, symmetry of shape, perfection of execution, diversity of form, harmony of relationships"? Consciousness, humankind's blessing and curse, gives us the ability to step back in order to observe the actions and processes of other creations and creatures. With this ability to observe comes a responsibility. For if it is possible to act as the overseer, then it is encumbent upon us to protect all. We should play the roles of benevolent custodians and not malevolent keepers.

Introduction

Nature is relationships in space.
Geometry defines relationships in space.
Art creates relationships in space.

> The essence of a painting lies in its construction of space.
>
> David Burnett
> 20th century art historian

Space, as we stated in Book 1, *Universal Patterns*, is the arena of all life's interactions. These are essentially three dimensional, having length, height, and width. Even the thinnest slice, or layer, has thickness, and, yet, when we work with the plane, we consider it to have only the two dimensions of length and width. It becomes an abstraction, an idea of something rather than the something itself. Here, in this book, we will devote our attention to the concept of the surface plane and to those actions we can perform to manipulate elements on it.

Mathematically the term *plane* is undefined and, yet, certain assumptions are made about it in order to work with, and within, it. In Euclidean geometry, it is understood to have infinite length and width, but no thickness. It can be thought of as being the flat surface that is obtained when space is sliced through. Although any number of two dimensional figures may be positioned and repositioned within the plane, the character of the plane itself remains neutral, static and unalterable.

For the visual artist, however, the plane is neither unbounded nor neutral, but becomes charged with energy and meaning, even by a single mark. Beyond having physical, geometric and visual properties, it functions as symbolic space as well. It becomes a metaphor for describing in two dimensions what is perceived as existing in three. For the artist, the understanding of both the nature and the function of the plane is of prime consideration.

Over time, human beings have altered their perceptions and conceptions of the plane, and their images have reflected these changes. Much of what has been created in the visual arts has been intended for the plane—a wall, a ceiling, a floor. How the individual artist accepts or rejects the plane's limitation of graphic depth reflects the particular mode of spatial thinking peculiar to that artist's cultural definition of visual structure. The artworks on this page and those that follow point out some of the differences.

Above:
Piet Mondrian. Diamond Painting in Red, Yellow, Blue. *Ca. 1921-1925. Diagonal: 1.428 x 1.423 m. Canvas on fiberboard. National Gallery of Art, Washington, D.C. Gift of Herbert and Nannette Rothschild.*

The ratio of the dimensions of the painting is approx. 1:1, a square. The concern is for the division of the surface area while suppressing depth.

Left:
Bukhara. School of Bihzad. The Nuptials of Mihr and Nawhid. *Persian painting: early 16th century (A.D. 1523). 19.3 x 12.2 cm. Courtesy of the Freer Gallery of Art, Smithsonian Institution, Washington, D.C.*

The ratio of the dimensions of this page of a manuscript is 1:1.6, approx. a Golden Rectangle. Multiple viewpoints are combined into a single image.

Right:
Li Ch'eng (Ying-ch'iu). Buddhist Temple Amid Clearing Mountain Peaks. *Chinese, ca. 940-967. Ink and slight color on silk (44" x 22"). The Nelson-Atkins Museum of Art, Kansas City, Missouri (Nelson Fund).*

The ratio of the dimensions of this work is 1:2, a √4 Rectangle. The image is drawn as if seen from a vantage point high above ground level.

Century, many of the rules have been called into question, manipulated, altered, and have been redefined.

There is an inherent tension between the actual surface, which is a bounded portion of the plane, and the sense of illusion of depth when marks are organized upon it. Great art resolves the tension between the surface and depth illusion by interweaving them. Shapes, seen as objects, especially in representational or naturalistic work, exist at various planar levels. Depending upon the particular art period, one or the other aspect will dominate an image. But, all image-making on a two dimensional surface must contend with this inherent contradiction. Too much depth without surface considerations results in "holes" that visually destroy the surface unity. Too much surface emphasis, and the work becomes more decorative, like patterned wallpaper. Western tradition has had the tendency to ignore patternmaking

The picture plane is essentially a gameboard on which the pieces are moved about according to the rules defined by a particular culture, period, or both. The aim of any artwork is to present a unified image of techniques, materials, subject matter, and concept, even when the elements might suggest contrast. In the Twentieth

as a viable concern for the construction of images. Rather, it favors unique views. A culture, such as that of Islam, on the other hand, has long repressed representation in favor of pattern which takes on a metaphysical meaning. Obviously, no one culture can do everything that is possible in the manipulation of a surface. So, unconscious cultural factors that are philosophical and spiritual enter into the game.

The following are some of the assumptions about the picture plane (surface) as found in the western art tradition. Once established the rules become an invisible cultural structure.

1 There is a viewer/participant standing at a particular orientation to the work that is being viewed.

2 The picture plane is considered flat, parallel to the viewer, and seen head on. If, however, the plane were to become concave or convex, then a different set of game rules would apply. We shall investigate some of these differences in Book 4, *Malleable Space.*

3 The plane is usually a rectangle, or a group of rectangles, which means the format shape contains right angles. The plane itself may be part of an architectural element, or a free standing panel, which still maintains the right angle relationship. This plane is considered as having only one surface to be manipulated. The obverse side is to the wall and considered visually non-existent. Elements usually do not flow from one side to the other. Thus, the work appears to be self-contained.

4 The rectangle is perceived

to be a transparent window out of which the motionless viewer looks onto a specific scene that is frozen in both space and time. The boundaries of the rectangle define the limits of the view. Movement is implied rather than actual. Everything happens "out there" in front of the window.

5 On this picture plane, there is a top to bottom and a left to right relationship. Each corner, each quadrant possesses a different power for holding the viewer's attention. This may be directly related to how a culture presents the written word, or it may

Below:
Marcel Duchamp, American, 1887-1968. Nude Descending a Staircase #2. *1912. Oil on canvas. 58" x 35". Philadelphia Museum of Art. Louise and Walter Arensberg Collection.*

The ratio of the dimensions of the painting is 1:1.6, approx. a Golden Rectangle. There is an implied sense of movement across the surface with a suppression of deep space.

very well be rooted in the perceptual mechanisms of all human beings. There are implied vertical and horizontal axes that act as points of balance. Shapes, having visual weight as opposed to actual physical weight, are organized about these axes. Forms plane, varying illusions of depth are suggested. The use of linear perspective defines the ways in which objects, especially architectural ones, recede and move away from the viewer, while atmospheric perspective creates illusions of depth through changes in the intensity of color and the clarity of detail.

7 All pictorial elements are presented through the use of a homogeneous material. It is through the techniques and skills of the artist that the illusion of actual materials and textures are created. However, the process of collage in the Twentieth Century has given the artist the freedom to juxtapose all kinds of elements, both actual and implied as well as to create visual contradictions.

In this book, again, our concern is to link the worlds of ideas and forms and to bring to them the harmony that springs from understanding the organization of the natural world. To that end, we shall explore how the universal patterns of Book 1 from the *Golden Relationship: Art, Math & Nature* series, can affect and manifest themselves in the *Surface Plane*.

This page:
William Michael Harnett, 1848-1892. After the Hunt. 1885. Oil on canvas, 71.5" x 48.5" (unframed). The Fine Arts Museums of San Francisco, Mildred Anna Williams Collection.

The ratio of the dimensions of the painting is 1:1.414, a √2 Rectangle. There is the illusion of three-dimensionality despite the suppression of depth. This is in the genre of trompe l'oeil painting.

Opposite page:
Raphael Sanzio. The Marriage of the Virgin. 1504. Oil on round-headed panel. 67" x 46". Pinacoteca di Brera, Milan, Italy.

The ratio of the dimensions of the work is 1:1.4, approx. a √2 Rectangle. Linear perspective is used to create illusions of deep space.

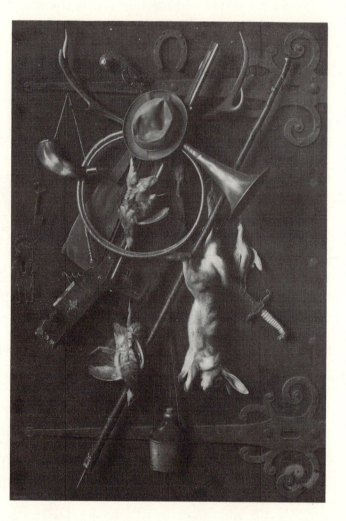

nearer to the central point are perceived as having less weight than those that are closer to the edges of the format shape. 6 Through the pictorial devices of overlapping, transparency (seeing one form through another), chiaroscuro (the manipulation of light and dark values), and changes in relative size and position of objects on the picture

1 The Grid

We cannot investigate Nature through either Mathematics or Art unless we set boundaries to our investigation. One way of assigning a structure to our exploration of the surface plane is through the use of a grid. A grid is a repeating pattern consisting of lines and/or circles superimposed on the plane. It is an intellectual tool suggested by such natural forms as cracking patterns, spider webs and cellular arrangements.

An artist has to train his responses more than other people. He has to be as disciplined as a mathematician. Discipline is not a restriction but an aid to freedom. It prepares an artist to choose his own limitations.

Wayne Thiebaud
20th C. American artist

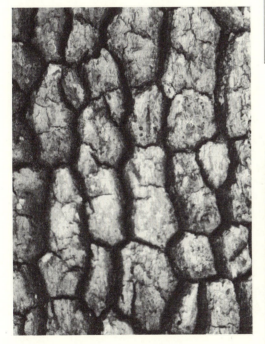

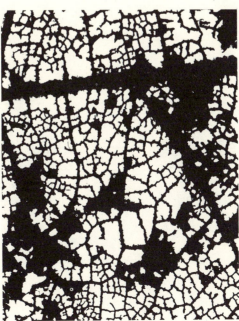

This page:
Forms suggestive of grid structures

Top:
Cracked asphalt paving

Far left:
Pine tree bark

Near left:
Leaf skeleton

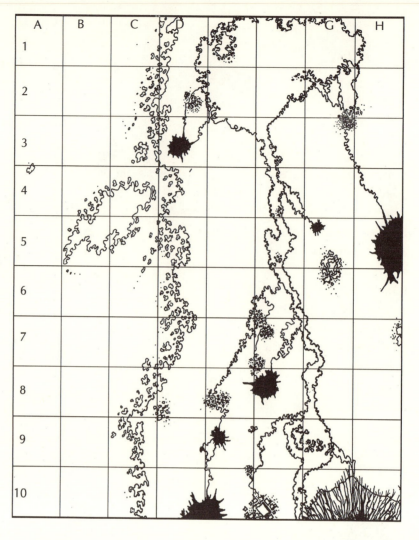

Right:
Fig. 1.1 On a map, the grid acts as a visual sieve to help locate particular places. Fantasy map drawing by Graham Boles.

Below:
Fig 1.2 In the template for cross stitch embroidery the grid functions as the structure. Edmund V. Gillon, Jr. Geometric Design and Ornament. *Courtesy of Dover Publications, Inc.*

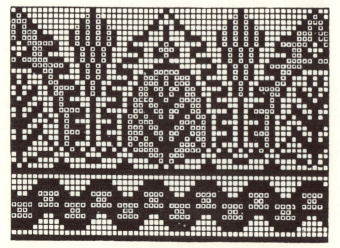

It allows us to set limits to the plane, not by enclosing a portion of it, but by dividing it into units so that information can be plotted onto these divisions. These units allow for the taming of seemingly unmanageable areas of the surface.

A grid can function either as a visual sieve (Fig. 1.1) or as a visual structure (Fig. 1.2). As the former, it becomes invisible and allows us to examine a section of ground while, as the latter, it insistently becomes the figure. But implied in both these aspects of the grid is the notion that some things are included in the visual field while other things are excluded.

In working with the surface plane as a two dimensional experience, there are two archetypal spatial systems. The first is one composed of straight lines, in which there are an infinite number of units that replicate themselves with regularity. The second is the concentric one, in which there is organization around a pole. Each has its distinct advantages but they also may be used in combination.

Straight Line Grids

A system of lines that consists of both parallels and perpendiculars yields a most convenient framework for spatial organization. It was used as early as ancient Egyptian times where body measurements determined the grid. The closed fist represented the size of the unit square.

In its simplest form the grid consists of equally spaced intersecting lines resulting in a planar subdivision of congruent squares with no obvious dominance of one direction over another. This basic structure may be manipulated in a variety of ways to produce other grids in which the lengths of the sides of the units have different ratios.

For those concerned with harmonic relationships, as was Jay Hambidge in working on his analysis of Greek artworks, a new set of grids is suggested. The Dynamic and Ø-Family Rectangles provide the initial units for these. They seem to suggest the inherent proportions in organic growth more readily than the square grid. These are reproduced in full in Appendix E, but should you wish to develop a grid, Constructions 1 and 2 are useful.

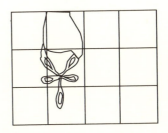

Left and below:
Fig. 1.3 The closed fist determined a unit for the square grid, and became a basic module for measurement.

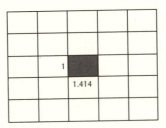

√2 Grid

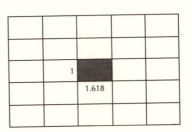

Golden Grid

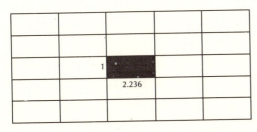

√5 Grid

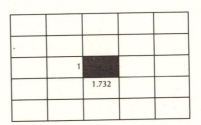

√3 Grid

Center and far left:
Fig. 1.4   Some examples of Dynamic and Ø-Family Grids. Here the unit width has been kept constant while the length varies. Remember the length of any line segment could be chosen to have unit measure.

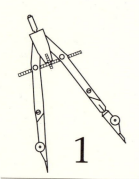

1

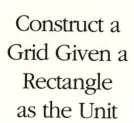

Construct a Grid Given a Rectangle as the Unit

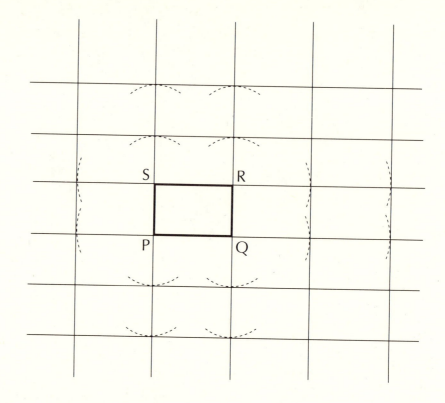

Given PQRS a Golden Rectangle

(We have used a Golden Rectangle for illustration, but any other type may be substituted.)

1. Extend the sides of PQRS as far as desired in either direction.

2. Open the compass to measure PQ, and mark off units in both directions on $\overleftrightarrow{PQ}$ and $\overleftrightarrow{SR}$, beginning at the vertices of PQRS.

3. Open the compass to measure PS, and mark off units in both directions on $\overleftrightarrow{PS}$ and $\overleftrightarrow{QR}$, beginning at the vertices of PQRS.

4. Connect points of intersection to draw lines parallel to the sides of PQRS.

The resulting grid is a Golden Grid.

Note: All directions are written as if the constructions are to be done by hand. In fact, they were executed on the computer. Different approaches are required when moving to the electronic tool. Primarily, one must think and work in layers. However, we have noticed that, in the initial stages, work requires time, attention to detail, and careful craftsmanship, whether the tool be a pencil or a mouse.

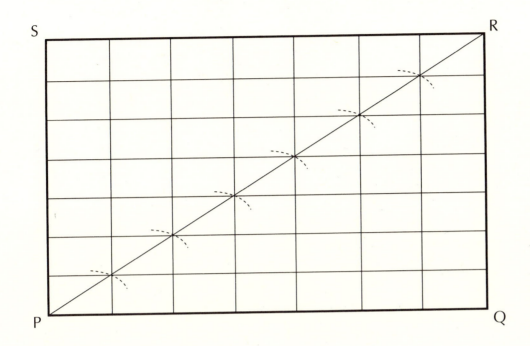

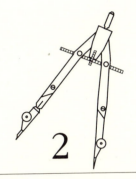

2

Subdivide a Rectangle into a Rectangular Grid of the Same Proportions

Given PQRS a Golden Rectangle*

(See Ch. 4, Book 1, Universal Patterns)

1. First decide how many units are wanted along the width and length of the rectangle. (For the sake of illustration, we will use 7, but any number may be chosen.)

2. Draw $\overline{PR}$ and subdivide it into 7 (or the number chosen in Step 1 if other than 7) congruent segments .*(See Construction 8, Book 1, Universal Patterns)*

3. Through each of the points along $\overline{PR}$, construct a line parallel to $\overleftrightarrow{SR}$. *(See Construction 7, Book 1, Universal Patterns.)*

4. Through each of the points of intersection along $\overline{PR}$, construct a line parallel to $\overleftrightarrow{RQ}$.

Now the rectangle is subdivided into a Golden Rectangle grid.

*We have used a Golden Rectangle for illustration, but any other type may be substituted.

Rene Descartes 1596 - 1650

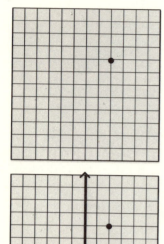

Given one of these rectangular grids, how can we describe the location of a particular point? Where do the concepts of left, right, up and down fit in? It is an essential limitation of the grid that these questions cannot be answered without additional information. The problem, however, was solved in the 17th Century by the French mathematician/philosopher Rene Descartes. It was he who realized that a center point could be designated by the superimposition of a perpendicular pair of axes on a rectangular grid. These axes were called the **x axis***, which defines the horizontal location, and the **y axis**, which defines the vertical. With the addition of this pair of axes we can distinguish left and right, up and down. However, we still cannot determine just how far from the center, in any direction, our location is.

It was Descartes' insight which allowed for the merger of algebra, which involves numbers, with geometry, which involves space. He assigned numerical values to points evenly spaced (determined by the grid) along the two axes. On the x axis, the portion to the right of zero was chosen to be the positive direction, and negative numbers were assigned to the left. On the y axis, the positive numbers were assigned above zero and the negatives below. Any point on the plane can then be named by an **ordered pair** of numbers, (x,y), which tells where the point is located, first with regard to the horizontal, then with regard to the vertical.

This incredibly simple solution is one of the real historical gems. It gave rise to an entirely new branch of mathematics called analytic geometry. It allowed for algebraic proofs of geometric theorems, as well as creating a graphic means of depicting algebraic equations.

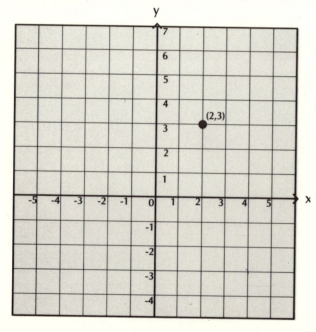

Above:
Fig. 1.5 *The use of the grid and Cartesian coordinate system enables us to locate a point in the plane. The point's location is able to be described only with the addition of the coordinate system.*

*All bold-faced terms are defined in the glossary.

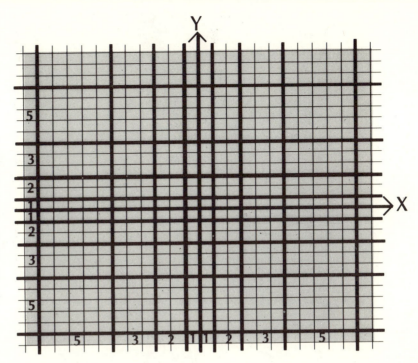

Grids need not be restricted to having units of the same dimensions, but can expand or shrink in accordance with a predetermined structure. For example, the Fibonacci sequence, 1, 1, 2, 3, 5, 8, ..., can be plotted along the coordinate axes and the result (Fig. 1.6), although rectangular, has units of unequal size.

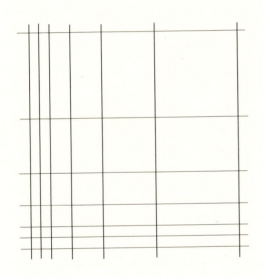

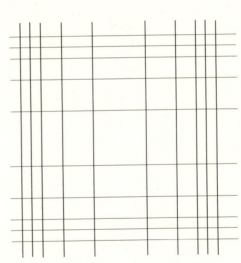

Above:
Fig. 1.6   Fibonacci subdivisions on the grid. Any point on the grid may be chosen as the "starting" point for the sequence. The choice is dependent upon the problem to be solved. Note that the numbers refer to the spaces between the lines and not the coordinates of the axes.

Left and center:
Fig. 1.7   Variations on the Fibonacci grid.

Practically speaking, the Cartesian plane makes life easier. Consider the way in which a position is located on a map of the world. In spite of the earth's curved surface, we isolate portions of it and consider them planar. We then impose a grid onto that portion and use it for reference. Distortion of land and water shapes occurs when we try to flatten the entire surface of the globe. However, this is a common practice among cartographers and the lines of latitude and longitude become the grid.

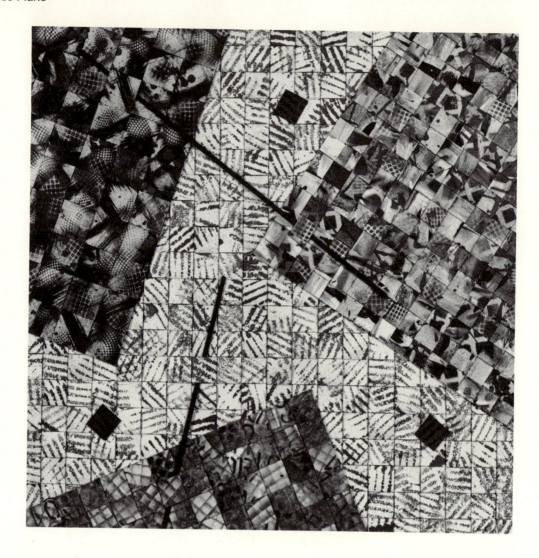

Above:
Kay Rosenberg. Printed Matter. 30" x 30", tarpaper, acrylic paint, wood, enamel paint.

My work uses the woven grid as its basic structure. This ordered system allows me to restructure and combine painted patterns that might otherwise be divergent. Choosing to maintain the element of randomness within these pieces, I am contrasting order and disorder in exploring the harmony of opposites. The framentation and displacement of repetitive forms creates a sense of movement. Space is defined and questioned as lines intersect with shapes on various planes. Light and shadow suggest depth and change. Geometric and organic forms compliment each other.

The grid sets the stage.

Kay Rosenberg
contemporary artist

Arrays of Points

Another way to look at the grid is through an array of points. If these are spaced equidistantly, only two possible patterns emerge. The first, where rows of points are placed directly under one another, frames the square grid we have already mentioned. If, however, the horizontal rows are adjusted so that the points in one lie between the points in the previous row, a new kind of straight line grid can be constructed. This array is said to be **isometric**. The units become equilateral triangles and, rather than two, there are three sets of parallel lines (Fig. 1.8).

In Appendix E, there are grid templates marked with these arrays. The square grid can be used as illustrated previously, or rows of points can be skipped to alter either the shape or size of the grid. If diagonal lines are used on this template, grids consisting of right triangles and parallelograms can be formed (Fig. 1.9).

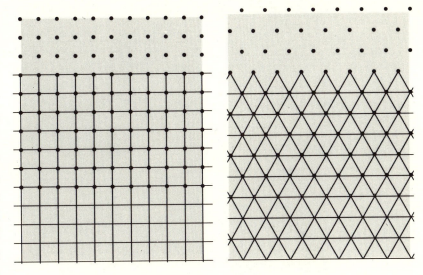

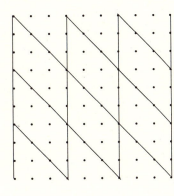

Above:
Fig. 1.8   The two basic dot arrays and their corresponding grids.

Left and below:
Fig. 1.9 A variety of grids drawn on the square array.

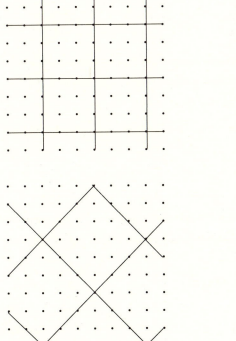

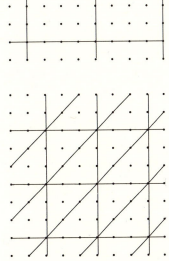

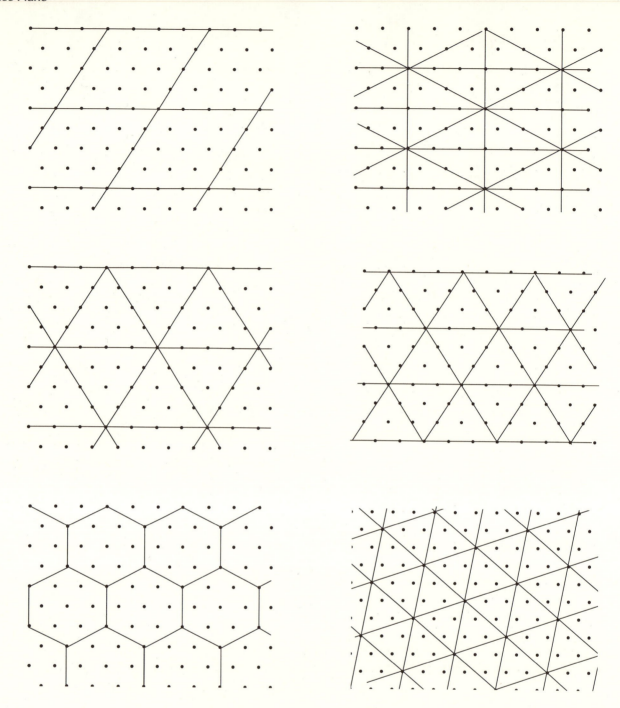

This page:
Fig. 1.10 Grids developed on the isometric array.

On the isometric array, the size of the basic equilateral triangle can be altered by skipping rows, or new patterns can be formed as illustrated in Fig. 1.10. Right triangular grids can be constructed on this template by using horizontal and vertical lines as well as diagonal ones. A wide variety of grids can be drawn on dot templates, but many cannot. Since the

Special Triangles and the Dynamic Rectangles and Parallelograms have harmonic proportions, they make particularly useful units for grids for design purposes. They cannot, with but a few exceptions, be drawn on the arrays. However, Constructions 1 through 4 show how to create your own.

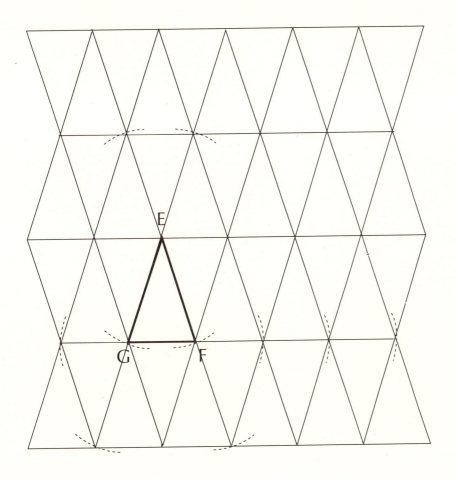

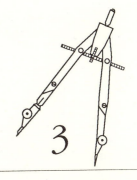

3

Construct a Triangular Grid Given the Unit

Given Golden Triangle* EFG (72°, 72°, 36°)

1. Extend $\overleftrightarrow{EF}$, $\overleftrightarrow{FG}$ and $\overleftrightarrow{EG}$ as far as the desired size of the grid dictates.

2. Open the compass to measure EF, and mark units of length EF on $\overleftrightarrow{EF}$ and $\overleftrightarrow{EG}$ repeatedly.

3. Through the points of intersection draw lines parallel to $\overleftrightarrow{FG}$.

4. Open the compass to measure FG, and mark units of length FG on $\overleftrightarrow{FG}$ repeatedly.

5. Connect the points of intersection of these two sets of parallel lines to form a new set of lines parallel to $\overleftrightarrow{EF}$ and $\overleftrightarrow{EG}$.

6. The remaining three lines can now be drawn in at the vertices E, F and G.

Now the system is a Golden Triangle grid in which the unit is Δ EFG.

*Any other type of triangle may be substituted.

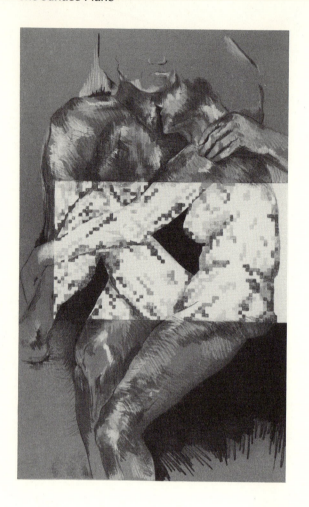

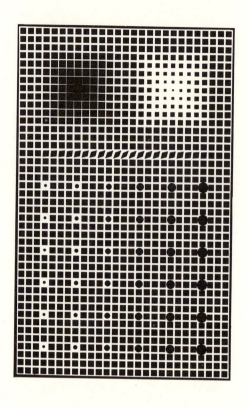

Above left:
Sharon Summers. Student
Work. Colored pencil on pa-
per.

Above right:
Victor Vasarely. Supernovae.
1908. Oil on canvas. The Tate
Gallery, London.

Lower right:
Robert Delaunay. Window on
the City No. 3. 1911-12: *Col-*
lection, Solomon R. Guggen-
heim Museum, New York.
Photo: Robert E. Mates.

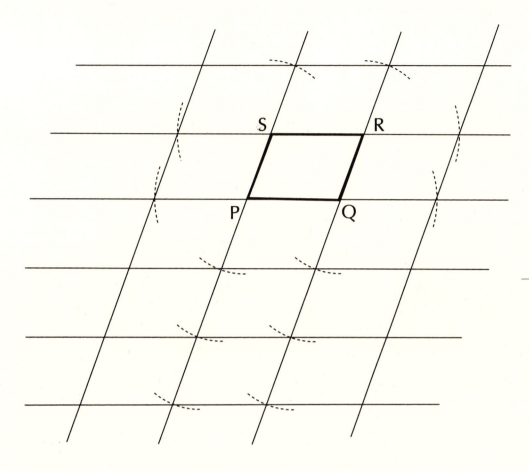

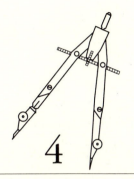

4

Construct a Grid Given a Parallelogram as the Unit

Given PQRS a √2̄ Parallelogram*

1. Extend the sides of PQRS as far as desired in either direction.

2. Complete the construction by following Steps 2 through 4 given in Construction 1.

Now the system is a grid whose unit is a √2̄ Parallelogram.

*Any other parallelogram may be substituted.

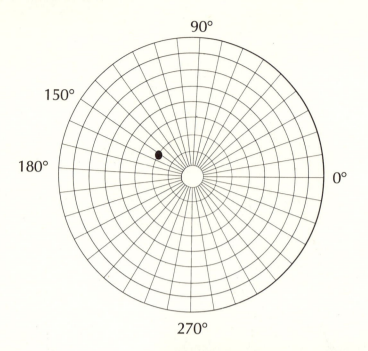

The Polar Grid

The other way in which a grid can be organized is around a center, or **pole**, from which vectors radiate at specified angles. Distances from the pole are determined along any vector cut by concentric circles. In its standard form, radii of the circles increase in **arithmetic progression**.

Again a location can be named using analytic geometry. This time the ordered pair is of the form (r, θ) where r names the distance from the pole (r standing for radius), and θ (the Greek letter theta) stands for the measure of the angle. For example, the point $(3, 150°)$ lies on the $150°$ vector 3 units from the pole. The mathematician uses this grid for visually describing certain kinds of equations, and more information about it is given in Appendix B.

Above:
Fig. 1.11   The Polar Grid.
The point (3, 150°) is
indicated.

In these photographs, both hu-
man-made and natural ele-
ments are suggestive of polar
grids.

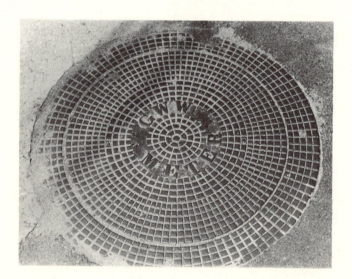

Center:
Cross section of a log
Left:
Manhole cover
Right:
Glass roof

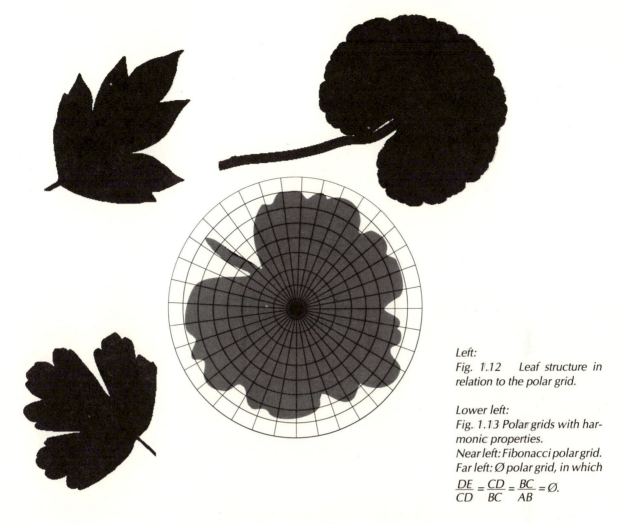

Left:
Fig. 1.12   Leaf structure in relation to the polar grid.

Lower left:
Fig. 1.13 Polar grids with harmonic properties.
Near left: Fibonacci polar grid.
Far left: Ø polar grid, in which
$$\frac{DE}{CD} = \frac{CD}{BC} = \frac{BC}{AB} = \emptyset.$$

Should we wish to analyze the growth structure of an essentially flat form which grows from a central point, such as a leaf, the polar grid provides a better model than the rectangular one. For design purposes we can alter the standard form and develop polar grids with more harmonic properties. Subdivisions of a line segment into Fibonacci or Ø proportions, as in Fig. 1.13, suggest divisions for the lengths of the radii of the concentric circles.

Circular Grids

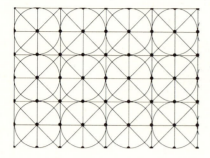
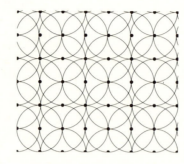

Another way that the plane can be subdivided by grids employs the use of circles. It differs from the polar grid in that the circles are either **tangent** or overlapping. These, too, can be constructed by using the grid templates as illustrated in Fig. 1.14. We will examine the uses of these more closely in Chapter 3 when we investigate the circle in the plane.

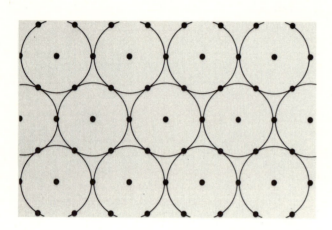

Above;
Fig. 1.14   Circular grids derived from dot arrays.

Right:
Hubcaps hanging on a fence hint at the notion of a circular grid. Unity is obtained through the repetition of circles; variety is achieved through the differences within each circle.

Uses of the Grid in Design

Both the mathematician and the artist look to the natural world, but toward different ends. In mathematics as well as art, elements are consciously organized into a unified whole. When happening upon a new landscape, the artist and the mathematician, walking side by side, might notice different aspects of the unfamiliar territory. Encountering a snake by the side of the river, the mathematician may be struck by the similarity of form, disregarding the scales on the snake and the ripples on the water. These, however, may be the very things that delight the artist.

The mathematician is interested in describing and explaining so that a situation can be generalized and repeated with some kind of predictability. The variations of the individual case are

purged and the simplicity of generalization is embraced. The artist, on the other hand, might become involved with sensuous variations and permutations on a theme which are then expressed in tangible form through a given medium. A response then moves from a purely abstract concept through to a personal physical representation.

Fig. 1.15 When natural forms are abstracted into graphic symbols, art, like mathematics, generalizes.

In much pattern making from cultures around the world, the grid structure becomes a constraining, but visible, part of a design. Some units are filled to become the figure while others are ignored and stay ground. Endless variety is possible even with the simplest grid. These patterns are used most often for basketry and textiles where the grid is evident in both the weave and the manner of decoration. In early pottery forms, much of the surface patterning was derived from textile structures where, initially, woven baskets were coated with clay to make them waterproof. When the vessels were dropped into the fire, the reed work was burned away, leaving the clay impression.

Left:
Fig. 1.16   Moire pattern constructed by overlapping acetate photocopies of two different grid structures.

An interesting phenomenon occurs when two or more grids are placed over one another and shifted slightly. The resulting pattern is called a **moire**. The one requirement needed to produce such a pattern is that the interacting layers have alternating solid and open regions. Although these types of patterns are easily created and, for a long time, have been exploited for use in making silk fabrics, contemporary science has found other advantages in their use. They can act as models for physical phenomena such as light, sound, and water waves, and can represent solutions to extremely complicated mathematical problems.

Upper right:
Ruth Trussell. Poster design.

Upper and lower left:
*Richard LaRue. Illustrations
for announcements of athletic
events.*

A powerful contemporary example of the grid is the computer screen. It is filled with an array of points called pixels which define the subtleties of the images. The greater the number of pixels, the finer the gradation of curve that can be suggested by the essentially right angle relationship of the grid. Given the added components of a drawing "tablet", a drawing instrument, and a method for reproducing an image, an individual has a rich field of design possibilities to explore. Any of the ideas presented heretofore can be integrated with the computer.

Problems

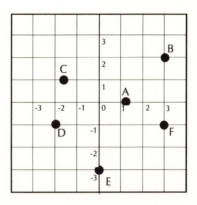

1 Name with an ordered pair of numbers each of the points indicated on the Cartesian plane in Fig. 1.17.

2 Name the coordinates that correspond to the tips of the leaf as it appears on the polar grid in Fig. 1.18.

3 From each of the two dot arrays (square and isometric) construct a grid that is not illustrated in the text.

4 Choose a Dynamic Parallelogram and construct a grid from it.

5 Construct a grid from the Triangle of Price (a right triangle whose sides measure 1, √Ø and Ø).

6 Find a simple image and overlay a grid of your choosing. Transfer the image to another grid which you will construct having a unit length three times that of the original.

7 Subdivide a rectangle of your choice into a grid composed of smaller similar rectangles.

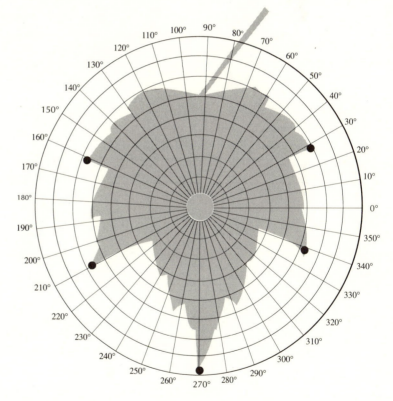

Above:
Fig. 1.18 Use with Problem 2.

8 Choose one of your favorite magazines or newspapers and see if you can determine the grid structure used in the layout. Modify the design using the information you have obtained from this chapter.

Projects

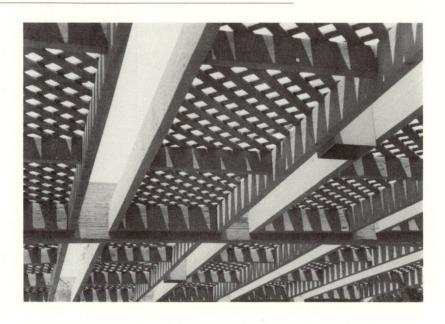

Above:
Patio lattice as seen from below reveals a grid structure. A practical structure becomes an abstract composition through the eye of the photographer.

1 Construct a square at least 8" x 8". Subdivide it with a square grid that has a minimum of 10 units on a side. You may draw the grid as if its lines have thickness, if you wish. Choose three colors that you find harmonious. Using color alone, create three different variations that change the appearance of the grid.

2 Using two found images of the same dimensions, subdivide and cut each with a grid pattern of your choice. Integrate the two in some fashion and reassemble.

3 Using one of the Dynamic or Harmonic grids discussed on page 3, create two different overall patterns with colors of your choice.

4 Choose any grid in this chapter and transform it in *one* of the following ways:

a. Using only black and white, develop an image or pattern that comes from filling in units on the grid.

b. Add color and decompose it in some way to form a new composition.

c. Color and cut in any manner and reassemble into an interesting three dimensional form.

5 Research the life and times of Rene Descartes. Using key facts from your investigation, create a crossword puzzle on a Cartesian coordinate system.

6 Using wires, strings, cords, threads, sticks, etc. create a wall relief structure based on one of the grids.

7 a. Using a grid overlay, take a natural form through a series of six transformations so that it evolves into a geometric shape that has dynamic or harmonic properties, *or*

b. Take a geometric shape that has dynamic or harmonic properties through a series of six transformations so that it evolves into a natural form.

8 Using a reproduction of an image from the history of Art, overlay a grid (straight line, polar or circular). On another surface of your choice, construct a similar grid using a size that yields a convenient unit for work. For each unit on the original image, assign a single color that best matches its color properties (hue, value, intensity). Use that color to fill in the corresponding unit on the other surface. Choose your own medium.

9 Using one of the Dynamic or Harmonic grids, and the concept of the reciprocal (*refer to Chapter 4 ,Book 1, Universal Patterns*), create an interesting overall pattern in black and white.

10 Overlay a polar grid on a leaf. Note the coordinates of key points on the leaf. Plot corresponding points on both the Fibonacci Polar grid and the Ø-Polar grid to create two transformations of the same leaf. Use the three variations as the basis for a composition with materials of your choice.

11 Use the structure of the grid to develop a Fibonacci plaid in the color harmony of your choice. (Transfer it to fabric and make a three-piece suit!)

12 Using black lines on a white ground, take one of the straight line grids and randomly curve various lines throughout the entire structure. Consciously alter the thickness and the type of line you use so that the viewer also becomes aware that the changes are a conscious choice.

*Above:
Claire Melanson. Student Work. Hand drawing scanned into the computer and re-worked.*

13 | *Aleatory Expression*

This project may be done in the materials of your choice. Do a study in which the following steps are taken:

a. Start with a Dynamic or a Ø-Family Rectangle. Subdivide it into a compatible grid.

b. Number all the units of the grid on a tracing paper overlay.

c. Number small slips of paper so that they correspond to the numbers on the grid.

d. Put the slips of numbered papers into a bag and shake them up.

e. Choose two colors, A and B. Draw out a single number at a time. Assign colors A and B, in alternating fashion, to each number until all the numbers are used up. Fill in the units on the grid with the colors indicated in this step.

f. Using this as a structure, with your chosen materials, manipulate or add as many personal elements as needed to "finish" the composition.

Below:
Richard Newman. Grid Play. *Photocollage.*

Further Reading

Arnheim. Rudolph. *The Power of the Center*. Berkeley: University of California Press, 1982.

Burton, David N. *The History of Mathematics*. Boston: Allyn and Bacon, Inc., 1985.

Doczi, Gyorgi. *The Power of Limits*. Boulder: Shambala Press, 1981.

Goossen, E.C. *Ellsworth Kelly*. New York: The Museum of Modern Art, 1973.

Hurlburt, Allen. *The Grid*. New York: Van Nostrand Reinhold Co., 1978.

Le Corbusier. *Le Modulor*. Cambridge: Harvard University Press, 1954.

Margolin, Victor. *Design Discourse History, Theory, Criticism*. Chicago: The University of Chicago Press, 1989.

Wade, David. *Geometric Patterns and Borders*. New York: Van Nostrand Reinhold, Co., 1982.

2 *Symmetry*

In all branches of science we try to discover generalizations about nature, and having discovered them we always ask why are they true. I don't mean why we believe they are true, but why they are true? Why is nature that way? ... there is simplicity, a beauty, that we are finding in the rules that govern matter that mirrors something that is built into the logical structure of the Universe at a very deep level.

Steven Weinberg
particle physicist

Throughout Book 1, *Universal Patterns*, we were primarily interested in examining harmonic spatial subdivisions in a recurring group of unique forms. In this chapter we will look at another "universal pattern", symmetry, which describes the ways those forms can be *positioned* in the spatial field.

The word symmetry, for many people, conjures up images of formality and stiffness; experiences that are artificially set and seemingly devoid of life. For others, it suggests the notions of balance, repose and proportion. None of these quite hits the mark. In Nature, when there is a state of total equilibrium there is likely to be symmetry. Yet, each element on this planet is prey to a multiplicity of forces bearing in on it in differing

degrees, so that pure symmetry is an ideal never to be realized for more than an instant. Scientists search for fundamental laws to explain the structure of matter. They attempt to construct simple comprehensive theories in order to define the workings of the universe. In theoretical physics today, mathematicians are using the idea of *supersymmetry* to clarify the actions and interactions of all basic particles. The notion of invariant operations helps scientists to speculate on elements that are not easily observable.

Above;
Several types of symmetry are illustrated in this architectural detail.

Yet, symmetry operations are as natural as breathing in and out. To understand the concept you need only to look at yourself in a mirror. What you see is a body in which one half appears to be identical to the other. The difference is designated as right and left. Look at your entire reflection. The three dimensional you is flipped into an image that exchanges left and right. Look down at your hands and notice that the left and right are mirror images of one another. Now, if you stand at the edge of a lake, the water acts as a mirror and you see yourself reflected upside down along the horizontal ground line. Move one step to either side and you have translated yourself in the plane. Take a few steps forward or backwards, stand on your hands, and your feet have performed the operation of glide reflection. Do a cartwheel and you have both rotated and translated yourself in space. Pivot on one foot like a ballerina and you have rotated about a point.

This page:
DeCat is descended from the house cat of the great French mathematician, Rene Descartes, from whom he took his name. Having lived in Massachusetts for many years, DeCat's name has not only been anglicized but also regionalized. His Boston accent is purrfect. Here, he does a dance of symmetry.

Above:
Claire Melanson. Student
Work. Computer scanned
image of an original drawing,
repeated and arranged
symmetrically.

Although natural, it is a very powerful concept, one which humankind has used in order to comprehend the workings of the Universe. Symmetry is a unifying, synthesizing, multidisciplinary principle uniting such widely diverse subjects as literature and geophysics, dance and crystallography, music and paleontology. It provides a model which is all encompassing and becomes a frame of reference by which we understand deviation. It can be used analytically to examine natural structures or synthetically as a way to create art.

In the major painting, *The School of Athens,* by the Italian Renaissance painter Raphael, symmetry is combined with intellectual discourse. This work contains layers of interdisciplinary meaning and unites the fields of art, mathematics and philosophy in the visualization of the search for Truth. The cast of characters comes from these three disciplines, and the drama is played out on an earthly plane.

The central figures, Plato on the left and Aristotle to the right, hold their respective great writings, *Timaeus* and *Ethics.* Their gestures highlight their philosophical differences. In the foreground on the left is Pythagoras. A student attempts to get his attention

Below:
Raphael. The School of Athens. 1510-11. Fresco. Stanza della Segnatura, Vatican Palace, Rome. Photograph courtesy of the Vatican.

Above:
Fig. 2.1 Symmetrical structure in Raphael's painting, The School of Athens. In this visual analysis, notice that the horizontal line connects Golden Cuts of the sides. The perspective center falls on the hand of Plato. A large triangle unifies the figures in the foreground with those in the row behind.

by holding a tablet showing a diagram of the theory of musical sounds. On the opposite side, Euclid, using a compass, explains a theorem to a group of students. The major Renaissance artist Michaelangelo is purported to be the isolated, seated central figure in the lower foreground, while the figure of Plato shares the features of the famed Leonardo. Raphael modestly painted himself second from the far right.

The strong sense of symmetry in the structure emphasizes the importance of the gathering, negating any sense of casualness. It is, however, the deviations from absolute symmetry that provide variety and interest within the unity. Movement and countermovement is suggested as one figure hurries in from the left and another exits on the right. Fig. 2.1 indicates key points, lines of action, and major subdivisions within the painting. Notice the hand of Plato in the center, anchoring the work.

This page:
The visual art of prehistory was developed on the uneven surfaces of rock walls without the framework of vertical and horizontal relationships.

All symmetry operations are spatial maneuvers. Thus, there are a limited number of moves to be made. These occur both in two dimensions, on the surface plane, and in the three dimensions of volumetric space. In this book we will deal only with the two dimensional moves, while in Book 3, *Structure in Space*, we will work with the third dimension.

In order to develop the concept of symmetry, there first has to be an underlying sense of vertical and horizontal, to which all other spatial movements orient themselves. It is subliminally felt and allows us to see where deviation from order occurs. In primeval art, space was perceived and expressed as multidirectional. No one direction was considered of prime importance. With the arrival of neolithic civilizations, however, verticality became the coordinating principle along with its corollary of horizontality. The grid, which we examined in the previous chapter, is a visible expression of a vertical-to-horizontal relationship. Adding the Cartesian coordinates, that relationship is given specificity. In this chapter these two concepts will be used to explore symmetry.

Symmetry Operations

In general, the operations of symmetry are performed unconsciously or occur naturally. Even though our everyday vocabulary lacks words that describe specific symmetry maneuvers, these operations are clearly defined mathematically. Symmetry operations are explored in Group Theory, invented by French mathematician, Evariste Galois at the beginning of the Nineteenth Century. It is a branch of higher mathematics that has applications in fields including and beyond the sciences. When mathematics defines the operations, we become more conscious of them, and awareness can lead to deliberate use.

In our explorations of symmetry operations we have selected, for illustration, a motif based on the Opah. It is a fish which Nature chose to design within a Golden Rectangle and we have chosen to stylize (Fig. 2.2). To further clarify the concepts, we have used the motif in conjunction with the Cartesian or Polar grids.

In the four standard symmetry operations discussed in this section, neither the size nor the shape of the motif is ever altered, and usually it is not overlapped. The operations involve only a repetition of the motif in various positions. It is this repetition that provides the basic element of unity in design.

Above:
Fig. 2.2   Stylization of the Opah.

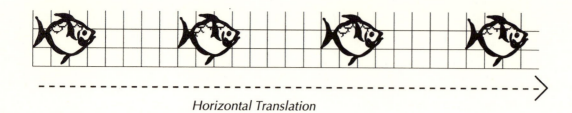

Horizontal Translation

Vertical Translation

Oblique Translation

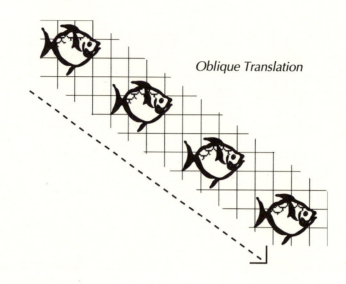

Translation

Translation is the sliding or gliding operation which changes the position of the motif without altering orientation or handedness. Any one of the moves in Fig. 2.3 qualifies as a translation. In each case the dotted line indicates the direction of the translation. The grid is provided as a reference for orientation.

Although any move which satisfies the restrictions aforementioned is a translation, in pattern making additional constraints are required. If the operation of translation is used, the space between the units must be a consideration. It may stay constant, as in the previous diagrams, or the spaces between the units may take on harmonic proportions. In Fig. 2.4 the spaces between the motifs are in Fibonacci progression.

Above:
Fig. 2.3 Translations of a motif in three directions.

Right:
Fig. 2.4 Translation of a motif with spacing between units in Fibonacci progression.

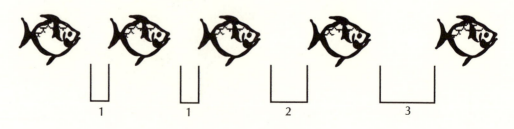

1   1   2   3

Rotation

Rotation is the turning of a motif about a point in either a clockwise or counterclockwise direction. The spatial orientation of the motif is changed, but the handedness is preserved. The motif is conceptually anchored to the point about which it is rotated, much like a pebble caught in the tread of a tire. It remains the same distance from the hub while constantly changing position as the wheel turns.

The pattern that is created depends upon the distance maintained from the central point and the angle through which the motif is rotated. Since this operation requires a pole and an angle, it is best illustrated using the polar grid.

The circle contains a total of 360°, so the number of degrees in the chosen angle must divide 360° without a remainder. Fig. 2.5 illustrates the difference between a rotation through 90° and one through 120°. In the first case, the motif is repeated four times, whereas in the second there are three repetitions.

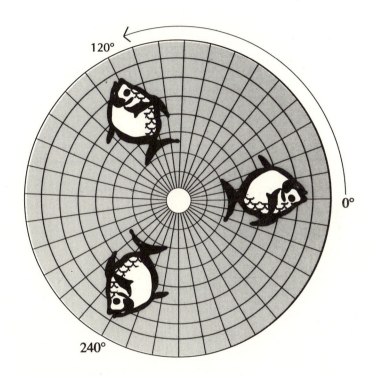

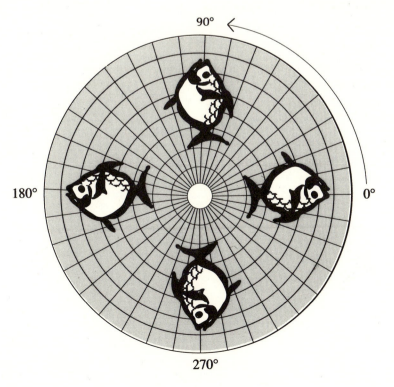

This Page:
*Fig. 2.5 Rotations through 120°
and 90°, respectively.*

Below:
Fig. 2.6 Angles of rotation required for particular numbers of repetitions of a motif.

Number of Repetitions	*Angle of Rotation*	*Positions of Radius Vectors Used*
2	180°	0°, 180°
3	120°	0°, 120°, 240°
4	90°	0°, 90°, 180°, 270°
5	72°	0°, 72°, 144°, 216°, 288°
6	60°	0°, 60°, 120°, 180°, 240°, 300°

Below:
Fig. 2.7 Polar grid with vectors at 10° intervals

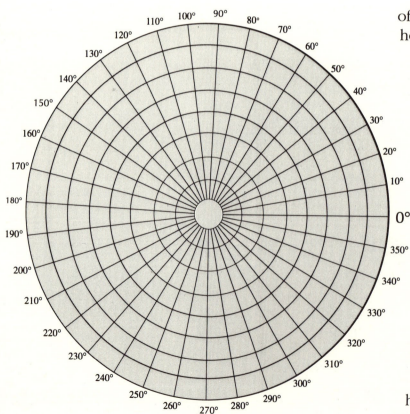

In general, to determine the size of the appropriate angle, first decide how many repetitions of the motif are required. If 360° is divided by that number the quotient will give the desired angle of rotation as illustrated in Fig. 2.6.

The motifs are then centered on the radius vectors named in the third column. Notice that each uses the 0° vector which is equivalent to the 360° vector. For design, any vector may be chosen as the starting point, or 0° position, providing that the others follow in sequence. Mathematicians say an angle is in standard position if the initial side (corresponding to 0°) lies on the horizontal vector to the right of the pole, as in Fig. 2.7.

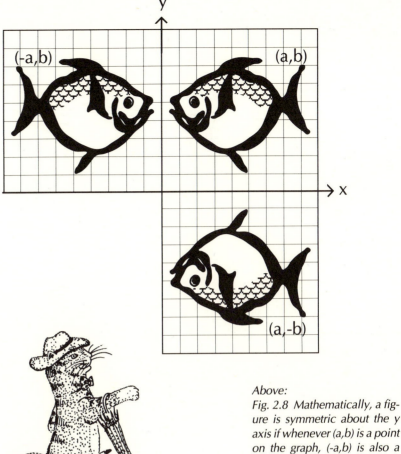

Reflection

Both translation and rotation operations can be achieved by sliding the motif in the plane, while **reflection** requires that the motif be pulled off the plane. It is the flipping up-and-over operation that changes both the position of the motif and its handedness in its relation to a line which we will call the **axis of reflection**. The figure in the two **half planes** are mirror images of each other. Because of this, the resultant image is sometimes said to have mirror symmetry.

Glide Reflection

Glide reflection is the combining of the operations of translation and reflection. The motif is first translated along an axis and is then reflected. Both the orientation and the handedness change. Muddy footprints on a freshly washed kitchen floor exhibit glide reflection (and lead to mother frustration).

Above:
Fig. 2.8 Mathematically, a figure is symmetric about the y axis if whenever (a,b) is a point on the graph, (-a,b) is also a point on the graph. In the illustration the fish that are facing each other are symmetric about the y axis. Symmetry about the x axis occurs if whenever (a,b) is a point on the graph (a,-b) is also a point on the graph. The two fish with ventral fins together are symmetric about the x axis.

Below:
Fig. 2.9 A glide reflection using the fish motif. The motif must be translated before it is reflected.

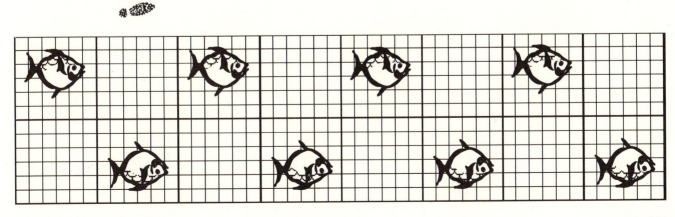

Below:
Fig. 2.10 Scaling combines translation with proportional changes in size.

Scaling

Not as familiar as other symmetry maneuvers, **scaling** is the concept that combines the operation of translation with a proportional change in the size of the motif. It is the only symmetry operation that allows for a change in the size of the motif. As with the single operation, added interest may be given to a design by considering the spaces between the motifs. The photocopier, or computer, is a most useful tool for making proportional changes, as it will increase or decrease an image by the same percentage each time.

Studio tip: When using a copier less distortion of shape and size occurs when the photocopier is reducing an image as opposed to enlarging it. Therefore, start large and work down to a smaller image.

Below:
Architectural detail exhibits the symmetry operation of reflection about a vertical axis.

Symmetry in Nature

All of the previous operations occur in Nature. Each kind of symmetry is a solution to a different type of environmental situation. In the evolution of form there is a logic to natural design. Everything is determined by practical need, that is survival, and form is the outcome of forces acting upon a given material in a given space.

In the plant and animal kingdoms, the essential push for survival depends upon adaptability to particular ecological niches. In all dynamic systems, forces are set against each other in order to reach equilibrium. However, in the natural world, so many forces play against one another that total equilibrium rarely occurs over extended periods of time. Symmetry is affected by position in space, the direction of gravity upon an object, and self motion. The needs for feeding, moving, and species reproduction are also factors that affect the form of an organism.

The two distinct types of environments, water and air, require different structural solutions to form. In the evolution of this planet, water was our first home. The earliest life forms floated free on the surface of the water and, because direction was unimportant, developed with spherical symmetry. As time passed and conditions changed, some creatures descended to the bottom of the oceans and became attached to the floor. In adapting to this sea environment, the directions up and down became important because of the need to maximize the amount of received sunlight. It was also important for these creatures, with their restricted movements, to develop a form that allowed them to capture food in the most efficient way.

Thus their bodies developed a cylindrical stalk topped by a radial series of spokelike feeding organs, much like the modern day sea anemone. Living forms like these that attach themselves to a surface, or those that move slowly across the sea bottom, such as starfish and sea urchins, distinguish up and down, developing a top and bottom, but have no need for differentiation of forward and backward, or right and left.

Radiating lines, as a construction device, have two useful attributes. They minimize the distance between the center and the outlying points, and they provide great scope for increasing the surface area of an organism. Radiating shapes also present a strong defense by giving the maximum numbers of spines around a body in order to discourage predators. Additionally, they give the impression of greater size. The plant kingdom, rich with radiating shapes, also takes advantage of these attributes, as most flowers have this form. Also, many plants grow leaves or branches that radiate directly from a stem base.

When creatures required more rapid locomotion, regardless of environment, a bilaterally symmetric structure evolved and a head developed to lead the way. These creatures, like those with radiate forms, have a top and bottom but, additionally, have a head and a tail. The sensory organs form a face. There is an advantage to having the eyes, ears, and nose close to the mouth where they are useful in the search for food. There is an equally great advantage in having them close to the brain. In these forms, asymmetry occurs as a secondary characteristic.

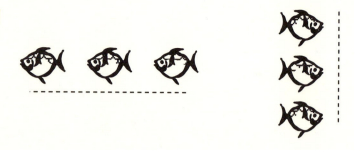

Symmetry Groups

When a unique motif is repeated in a systematic fashion, the result is a pattern. Symmetry deals with repetitions of different kinds. If the repetitions involve movement around a pole, the resulting collection of patterns is called the **point groups**. The **line groups** occur when the repetitions move in a single direction, and the **plane groups** result from repetitions in two directions so as to fill the plane.

Before discussing the groups in detail, some diagrammatic aids will help to clarify and make the concepts visual. A dotted line will always indicate the direction of a translation (see Fig. 2.11). A solid line indicates an axis of reflection. We have already discussed this concept in relation to the operation of reflection. However, we want to emphasize the fact that whenever an axis of reflection is used each image on *either* side of the line is reflected to the other side (Fig. 2.12). Finally, when a motif is used with various operations, the resulting pattern element becomes a **unit** which is used in multiples to produce a symmetry group. Shading will be used to define the unit (see Fig. 2.13).

Above:
Fig. 2.11 Dotted lines indicate the direction of translation.

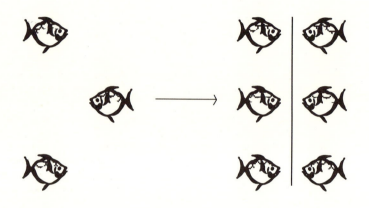

Center:
Fig. 2.12 The figure on the left is transformed to the one on the right when an axis of reflection is added.

Below:
Fig. 2.13 The shaded portion represents the unit, which is translated to create the pattern.

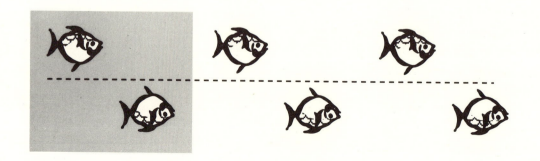

Point Groups

The point groups are patterns which are formed by the operations of rotation and reflection with respect to a pole or axis through the pole. For the mathematician, there are an infinite number of point groups since the physicality of these is not a concern. For the artist, however, constraints of area and size of motifs limit the number of point groups that are feasible for design. We will confine our discussion to these lower order groups.

In developing this section we, too, found constraints of space a problem with regard to our chosen fish motif. It was too chubby to rotate about a pole easily. We, therefore, developed a more angular motif based on the Golden Triangle (36°, 72°, 72°) and its subdivisions (Fig. 2.14).

You will notice that although the motif has harmonic proportions, it contains no inherent symmetry. It is, therefore, an asymmetrical motif, and as such is the lowest order point group. One of the determinants of a point group is that the entire group may be rotated about a point, called the **rotocenter**, through an angle less than or equal to 360°, so that it coincides with itself. An asymmetrical motif must be rotated a full 360° to coincide with itself. This group is called P1 (Fig. 2.14).

If we introduce an axis of reflection, our motif creates a unit that has bilateral symmetry. Where the axis is placed is the choice of the designer. In each of the illustrations in Fig. 2.15, the motif is the same, but because of the placement of the axis, the unit is different.

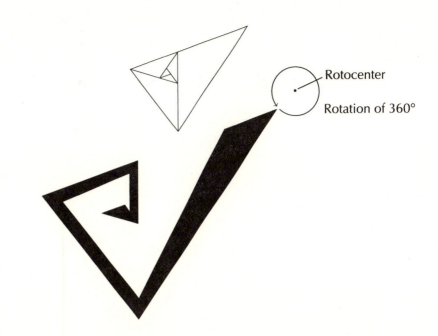

A unit with bilateral symmetry still requires a full turn about the rotocenter to coincide with itself, and therefore is related to group P1. We shall call it P1' (P-one-prime.)

Above:
Fig. 2.14 An asymmetrical motif created from the subdivisions of the Golden Triangle exemplifies point group P1.

Below:
Fig. 2.15 Point Group P1' is represented by an asymmetric motif reflected about an axis.

Rotocenter

Rotation of 360°

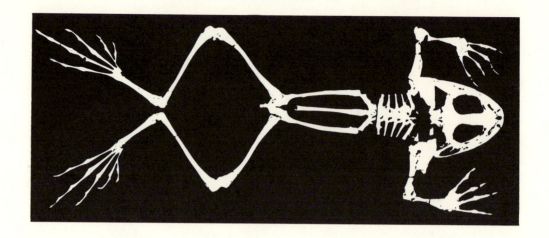

To obtain group P2, the motif is rotated a half turn, or 180°, as in Fig. 2.16. When an axis of reflection is introduced, as in Fig. 2.17, group P2' is formed. Here the original motif is reflected to produce a unit. This unit (the shaded portion) is, in turn, rotated 180°, which gives rise to a second axis of reflection indicated by $\overleftrightarrow{AB}$.

Group P1' has a single axis of reflection, P2' has two axes of reflection,

and the chart on the following page shows how the pattern extends through group P6'. We have chosen to designate as **base groups** the point groups obtained solely by rotation, and as **prime groups** those obtained by reflection *and* rotation. The metamorphosis from base to prime is shown for each of the members of these groups.

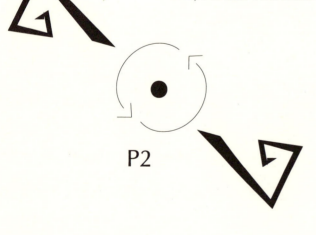

P2

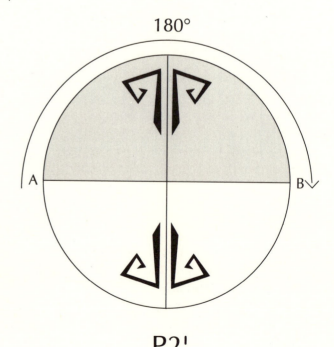

180°

A   B

P2'

Above:
Fig. 2.16 A half-turn of the motif results in group P2 .

Right:
Fig. 2.17 The motif is reflected to form a unit. The unit, shaded, is rotated 180° to produce a P2' group.

Base Group   With an axis of reflection   Prime group

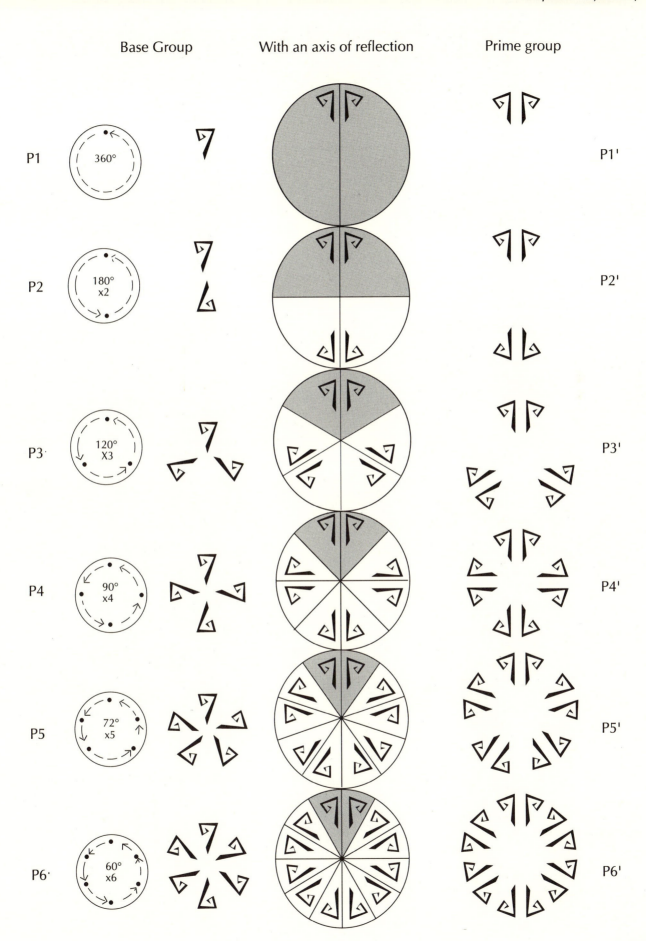

P1   360°   P1'

P2   180°   P2'
      x2

P3·   120°   P3'
      X3

P4   90°   P4'
      x4

P5   72°   P5'
      x5

P6·   60°   P6'
      x6

This page:
In the organic environment, prime groups are highly visible. The garden provides us with a great number of examples beyond the group P1', so evident in higher form skeletons.

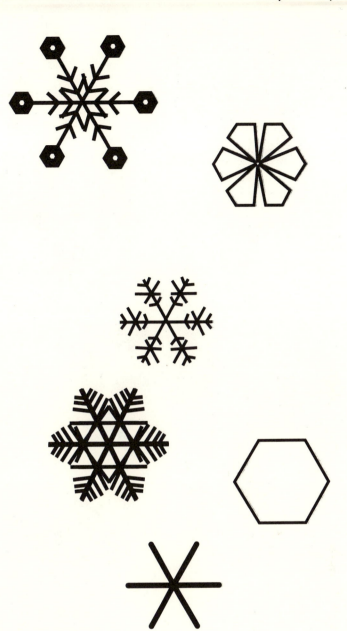

Nature severely restricts the kinds of point groups used to create inorganic forms. Crystals appear in groups of orders 2, 3, 4, and 6. Probably the most familiar example is the snowflake, always of order P6'.

Most of the major research done on the study of snowflakes was undertaken by a Vermont dairy farmer, William Bentley, a man with no formal education but with a love of the natural world and an absolute passion for snowflakes. Beginning at age 19, in 1885, he spent forty-seven winters in an unheated barn with primitive equipment for photographing and documenting 5,381 of these natural gems. It is a source of fascination that despite the similarity of structure of all snowflakes, no two are ever alike since subtleties of atmosphere influence how they are formed.

The hexagonal form of the snow crystal is a direct result of the shape of the water molecule. There are only nine variations on the basic structure, but each of these is prey to a myriad of minute variations due to natural conditions at the time of formation.

This page:
Fig. 2.18   The nine variations on snowflake structure.

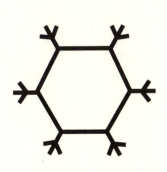

The point groups shown on the chart on page 45 are some of the most useful for design because of their simplicity and clarity. The artist-designer, however, is by no means limited to these lower order groups. For example, the stained glass Rose windows of Gothic cathedrals were often of the order P12', believed to be related to the life-giving properties of light. More information on Rose windows will be given in Chapter 3.

Top to bottom:
Japanese crests, P2, P3, P6. Fumie Adachi. Japanese Design Motifs. *Courtesy of Dover Publications, Inc.*

Geometric, P12 and Floral, P2. Jorge Enciso. Designs from Pre-Columbian Mexico. *Courtesy of Dover Publications, Inc.*

Floral, P4'. W. and G. Audsley. Designs and Patterns from Historic Ornament. *Courtesy of Dover Publications, Inc.*

Lace doily, P4.

One of the interesting ways to observe the point groups is through a kaleidoscope. This device consists of a cylinder that has an eyehole at one end, a flat piece of glass at the other, and two mirrors set in at particular angles within the body of the cylinder. As it rotates, the images seen through the eyepiece form the interlocking patterns of a particular point group.

An alternative to constructing such a device is to apply the concept to the plane by using a wedge shaped (**sector**) pattern for the unit, which is then taken through the proper number of rotations to complete the full image. The pattern can be used in different ways. It can be thought of as a window to be placed over found images which abound in magazines and the like. Sometimes it is desirable to simplify the color or line before repeating it in rotation. Or you can create a design directly within the wedge shaped piece and then take it through its rotations.

This page:
Fig. 2.19 Kaleidoscopic designs using a sector with an angle of 72 °.

> *. . . there is structure underlying all human behavior, that this structure can be discovered by orderly analysis, that this structure has meaning, and that there are as many kinds of structure as there are kinds of behavior.*
>
> Dorothy K. Washburn
> anthropologist

The Seven Line Groups

Linear patterns, sometimes called bands, borders, or friezes, have been widely used throughout time and cultures. Today, social scientists frequently use symmetry classifications for analysis of decorated objects. This enables them to classify things by similarity. It also simplifies the demonstration of cultural connections and differences, since decorated objects will show a culture's preference for a particular aspect of symmetry, shape and subject matter.

Linear patterns have a definite width but extend in either direction, controlled only by the character of the total design and the surface to be covered. A band is the simplest pattern using the essential operation of translation.

A unit may be created by using a vertical or horizontal axis of symmetry, a rotation of 180°, a glide reflection, or a combination of any two, or all three. And, yet, there are only seven possible band variations in terms of the placement of the motif within the unit. In the choice of motif, however, the possibilities are limitless, so that within these very tight restrictions it is possible to create with infinite variation.

The seven possibilities, using the Golden Triangle motif, appear in the chart on page 52. We have named the line groups with a capitol L followed by a shorthand notation describing the operations needed. The first group is simply a translation of the motif and is, therefore, named LT. The others require operations on the motif to produce the unit, and it is these which are described in the name of the group.

The square dot grid is used to show the structure of each of the Line Groups. The shaded area is the unit that is ultimately translated in order to create the pattern. The distance between units is an arbitrary choice but, once chosen, stays fixed. A variety of inventions culled from the history of design demonstrate each type of group.

Choice of the motif can alter the complexity of the design, and even more variation can be achieved with the addition of color. Although the patterns in Fig. 2.20 are highly stylized, one can see that the source of the motif is the natural world purged of all incidental information, abstracting out only the essentials of form. It is to the credit of the human mind that, when confronted with diversity, we are able to see similarity, finding common elements of unity. Line groups can be suggested by the natural world through observation of the phenomena that surround us all.

Above:
Claude Monet. The Poplars. *32.25" x 32.125". The Metropolitan Museum of Art, H. O. Havemeyer Collection. (29.100.110).*
This painting suggests a line group of order LRH.

51

The Seven Line Groups

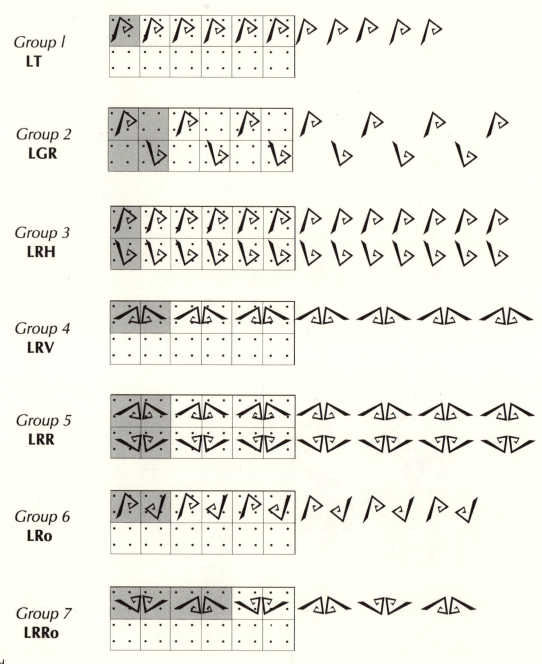

Group 1 **LT**

Group 2 **LGR**

Group 3 **LRH**

Group 4 **LRV**

Group 5 **LRR**

Group 6 **LRo**

Group 7 **LRRo**

Legend
T translation
R reflection
GR glide reflection
RH reflection about a horizontal axis
RV reflection about a vertical axis
Ro 180° rotation

Cultural Connections

LT

LGR

LRH

LRV

LRR

LRo

LRRo

This page:
Fig. 2.20 Line groups from the history of ornament.

Images are taken from these books, courtesy of Dover Publications, Inc:

Audsley, Designs and Patterns from Historic Ornament.

Enciso, Design Motifs of Ancient Mexico.

Gerspach, Coptic Textile Designs.

Naylor, Authentic Indian Designs.

The Seventeen Plane Groups

Pattern is the controlled repetition of a motif arranged in a particular manner with attention paid to the space intervals as well as the form units. The seventeen plane, or wallpaper, groups involve abstracting the essentials of pattern in two dimensional Euclidean space. Patterns of this type abound in the constructed environments of humans where surfaces call out for decoration.

Islamic artists discovered and used all the plane groups, the evidence of which can still be seen in the Moorish palace of the Alhambra in Granada, Spain, built during the 14th Century. As the Parthenon is to Greek culture, so this palace is the epitome of Islamic architecture and design. The art involved itself with abstraction because religious scripture forbade the use of images of the human form. The geometric forms became symbols of cosmic connections where the finite suggested the infinite. The plaster coated wall surfaces are covered with intricate patterns combining arabesque, calligraphic, and floral motifs in a seemingly never ending array.

It should be noted that there is a difference between how the mathematician and the artist consider the plane groups. For the former, the essential attribute is that of the relationship of a "motif" to a particular rotocenter or axis of symmetry. The motif is chosen as an indicator, not for its intrinsic visual properties. The pattern is used *only* as a diagram of orientation. At some point, even the motif gives, and is replaced by number, becoming a totally abstract concept. The mathematician looks at patterns to find different symmetry operations for the purpose of classifying them. The interest of the artist lies in creating the patterns. Therefore, for the latter, the form of the motif, the placement of it, and the spaces between motifs are all to be considered as part of the overall design.

One of the fascinating aspects of this area of investigation is the way in which complexity builds logically and elegantly from simple beginnings. One way to cover the surface plane is to create an original unit using point or line group operations, and then to

translate that unit in two nonparallel directions the required number of times. Since the space surrounding both the motif and the unit is critical and must be kept constant to form an overall pattern, the units are set within implied geometric frames. Once determined, the orientation of each unit to its frame in the pattern is always the same.

The following pages provide analyses of these seventeen plane groups using basic units and grids. With square and isometric dot grids, all of the necessary framework can be obtained easily. The dots are useful for positioning the motif to obtain the unit within the frame. For each one of the patterns we have shown how the motif creates the basic unit and then how that unit is translated in the plane. For groups 1 through 12, the diagrams most often show a unit consisting of a square or rectangle. This type produces a pattern in which rows and columns are at right angles. However, the unit is in fact a parallelogram which, when skewed, produces a sense of diagonal motion. Skewing changes the visual appearance but not the essential underlying structure. In groups 15 through

17, the unit may be either a hexagon or a parallelogram, depending on the spacing between elements.

As with the line groups, the name describes the operations required to create the unit. The PL in the name refers to the fact that the pattern belongs to a plane group. When followed by a numbered P notation, the essential unit is one of the point groups. If the unit is derived by using other symmetry operations, the abbreviations are the same as those used for the line groups.

You will notice that while the diagram indicates the orientation of the motifs and units, the various examples at first glance do not always look like the diagram. The differences lie in where the motifs are placed. The motifs were abstracted from natural images, and the patterns, generated by multiplying and shifting, have a very different feel from the single element. So that you may better understand the concept, a shaded region is overlaid on a particular pattern to identify its essential type. This region may or may not include an entire unit. It is dependent upon the proximity of the units when translated.

Group 1
PLP1

A simple translation of an asymmetric motif in two non-parallel directions.

With a rhombic unit

Note: The basic unit is a parallelogram of which the square, rhombus and rectangle are specific types.

With a square unit

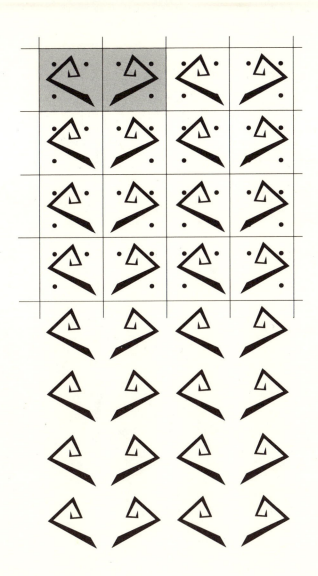

Group 2
PLP1'

A translation of a reflected asymmetric motif.

Remember: The triangular motif at the top of the page is used only as an indicator of the pattern structure. The pattern at the bottom of each page is based on that structure.

Notice that when units are placed very close to each other, the shaded region may contain parts of several units, but it is that region that will generate the entire pattern.

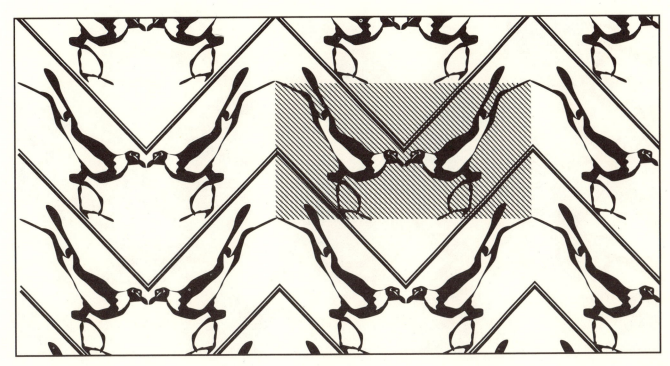

Group 3
PLGR

A translation of an asymmetric motif which is glide reflected.

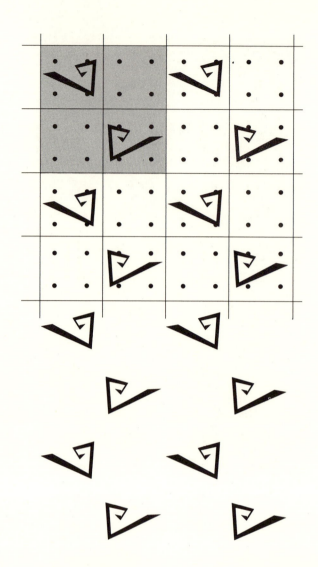

Notice that although the appearance changes, the essential character of the structure is the same given a full or partial glide of the motif.

Group 4
PLP1'GR

The motif is first created by a simple reflection (P1) and then is glide reflected.

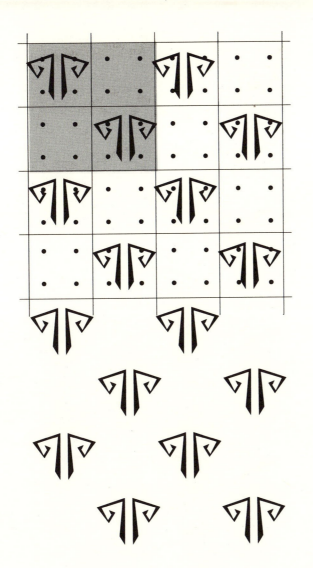

Note: a partial glide results in an overlapping of the motif. Also, even though the patterns on these two pages result from a reflection about a vertical axis, the reflection could be about a horizontal one as well.

Group 5
PLP2

The unit is created by a rotation of an asymmetric motif (P2).

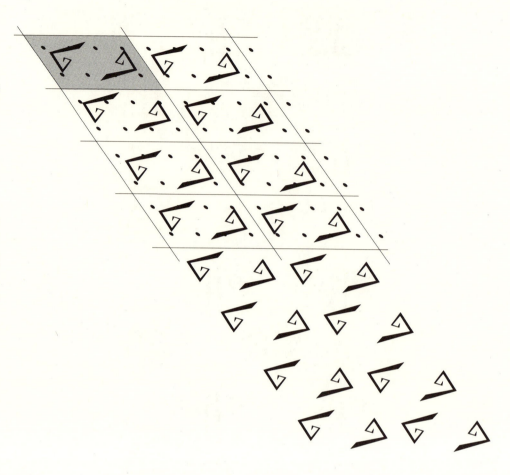

Note: the rhom-
bic unit could
be a rectangle.

Group 6
PLP2GR

A rotation of an asymmetric motif (P2) is glide reflected to create the unit.

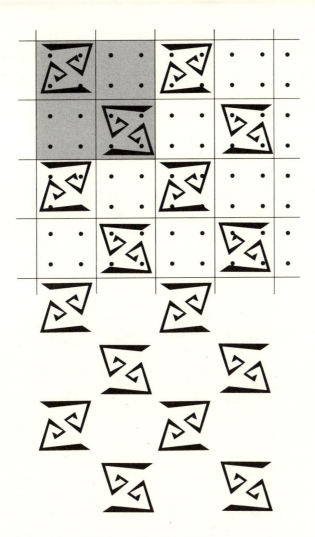

Group 7
PLP2'

A simple translation of the Point Group P2'.

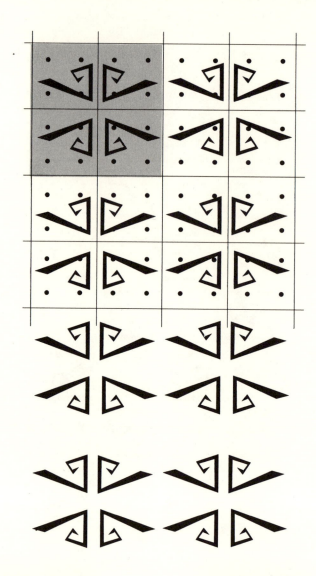

In this pattern, the physical structure is Group PLP2', but the color strucure is Group PLP2 (a rotation of a pair of the black and white squares used as a single motif).

Group 8
PLP2R

The asymmetric motif is first rotated (P2) and then reflected to create the unit.

Group 9
PLP2RR

A vertical and horizontal reflection of the Point Group P2 creates the unit.

Group 10
PLP4

*A translation of the Point Group
P4.*

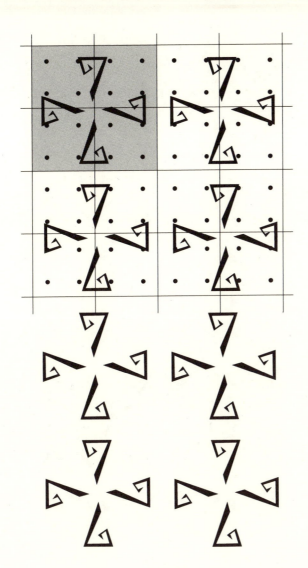

Group 11
PLP4'

A translation of the Point Group P4'.

Group 12
PLP4RR

A vertical and horizontal re-flection of the Point Group P4 creates the unit.

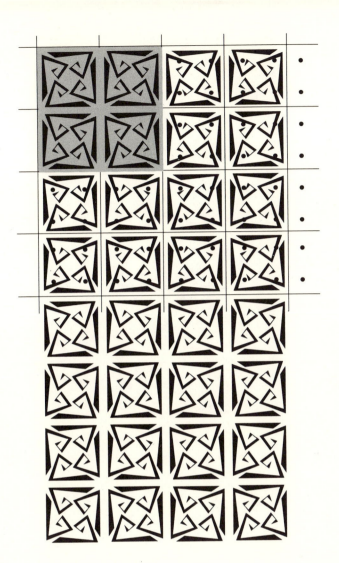

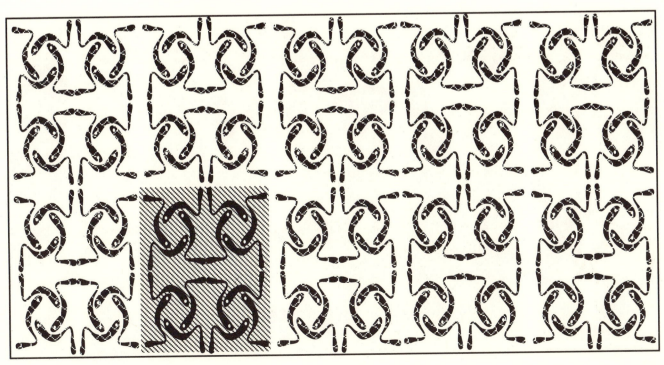

Group 13
PLP3

A translation of the Point Group P3

Note: The differ-
ence in pattern
depends upon the
placement of the
point group with-
in the rhombic
unit. When motifs
are pushed tightly
together, the shaded
region may not
contain the entire
point group.

Group 14
PLP3R

The unit is a reflection of the Point Group P3.

Group 15
PLP3'

A translation in two nonparallel directions of an hexagonal unit consisting of Point Group P3'.

Notice how the spacing between units determines whether the shaded region will be a hexagon or a parallelogram.

Group 16
PLP6

A translation of an hexagonal unit consisting of Point Group P6.

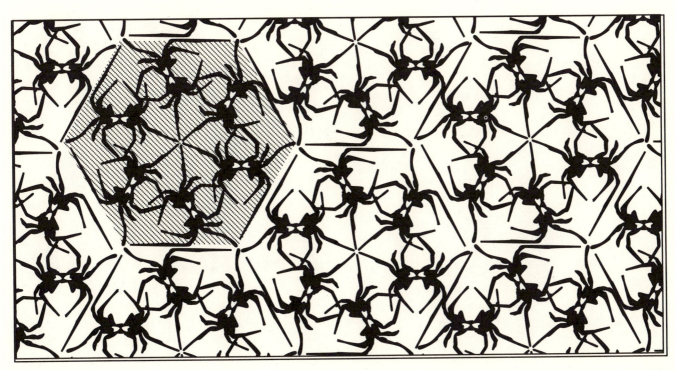

Group 17
PLP6'

A translation of an hexagonal unit consisting of Point Group P6'.

Above:
This natural structure hints at a
line group of order LGR.

In the previous examples, the idea and form of pattern is compressed into a relatively small spatial field so that the viewer immediately understands what is happening. Psychologically, perceiving pattern is reassuring. It suggests that there is order in the universe despite the obvious appearance of the random. The natural world is system upon system of recurring patterns. Therefore, our being sensitive to relationships within patterns, as well as the relationship of pattern to pattern, is crucial if our humanly constructed world is to parallel the structure of the natural one. Design is a way to sensitize people to the beauty of connections.

You will notice that the patterns we have examined are artificial constructs. Some are based on sources from the natural world and are reminiscent of experiences from it. But, they are highly stylized and simplified, consciously creating an aesthetic effect as the essential ingredient of the experience. In the natural world, beauty is a by-product. There is no creating of beauty for beauty's sake or art for art's sake, both of which are human activities.

In order to perceive a pattern, the viewer must be able to have some visual and intellectual distance from the subject. In the natural world, understanding may require the distance of time or the distance of tools, such as microscopes and telescopes, to aid in finding and then defining a pattern. Both mathematicians and artists find the search for and the creation of pattern emotionally and spiritually rewarding.

Problems

Above:
Architectural detail, P3'.

[3] Create a simple, but interesting, asymmetric line motif, and do the following:

a. Through the operations of rotation and reflection, create three variations in pattern using one of the point groups.

b. Through the operations of translation, rotation and/or reflection, create three variations in pattern using one of the line groups.

c. Through the operations of translation, rotation and/or reflection, create three variations in pattern using one of the plane groups.

[1] Choose a single number or letter of the alphabet that has neither rotational nor bilateral symmetry. Use it as a motif to demonstrate your understanding of the symmetry operations of rotation, translation, reflection, glide reflection and scaling.

[2] Demonstrate a translation wherein your motif is repeated at least five times in each of the following ways:

a. with a constant space between units,

b. with the space between units in Ø proportion *(see Construction 10, Book 1, Universal Patterns),*

c. with the space between units in Fibonacci progression *(see Construction 36, Book 1, Universal Patterns).*

Note: A computer or photocopier would be helpful here, or use tracing paper or a stencil.

[4] Choose two images from the history of art, one with symmetry and one asymmetrical. Analyze the structures using the concept of an armature *(see Chapter 4, Book 1, Universal Patterns).* Write about the differences you perceive.

[5] Find three examples of wallpaper, fabric or wrapping paper and determine which plane group is illustrated. Use tracing paper overlays to show your analyses.

Projects

1 Design an asymmetric motif using an harmonic armature of one of the Dynamic or Ø-Family Rectangles (see Chapter 4, Book I, Universal Patterns). For convenience of construction, let the width of the rectangle measure no less than six inches. Using the operation of translation (you may or may not use scaling as well) take your motif through a series of color changes from black and white to gray to a monochromatic color structure.

2 Design a motif that fits the requirements of the first part of the previous project. Use the photocopier to reduce and then replicate the motif. Create a border design using *one* of the seven line groups. Use your choice of black and white or color.

3 Design a motif as in Project 1 and create a wallpaper design using one of the seventeen plane groups.

4 Use the initials of your name in an interesting type face to create *one* of the following:
   a. a point group,
   b. a line group,
   c. a plane group.
Pay attention to aesthetic components and craftsmanship.

5 Using a 7 unit by 7 unit grid, one of the various harmonic ones given in Chapter 1, create a motif for an overall pattern using all seven line groups. Use the materials and colors of your choice.

6 Cut a sector from a magazine page suitable for a kaleidoscopic design. Trace the original as many times as needed to complete the pattern structure. Using the original as the first unit, take the remaining ones through a series of tone or color changes so that the finished design contains elements of both symmetry and asymmetry.

Below:
Leslie Miller. Original paper-cut. 1984. Newton Center, MA. Point group P4'.

7 Use folded tissue paper to design and execute a progression of point groups of orders P2', P4', P8'. Choose a color structure and take the groups through a color progression as well. Mount on illustration board using an appropriate adhesive.

8 Create an initial motif using a natural form as inspiration but designing it within a Dynamic Rectangle, Ø-Family Rectangle or Parallelogram variation. Develop all seven symmetry line groups using this motif. Choose *one* of these to develop into a border design for an 8.5" x 11" piece of stationery. Use your choice of black and white or color.

9 Begin as in Project 8. Choose one line group to develop into a composition that moves from black and white through grays into a particular color structure.

10 Using one or more of the 17 plane group patterns, create and execute a design for a kimono garment using the diagram in Fig. 2.21. It may be done in paper or fabric, small scale or actual size.

11 Develop a stylized motif based on a natural form. Choose one of the plane groups and create an overall pattern. Finish the work in materials and colors of your choice.

12 Transform a plain brown paper bag using a rubber stamp and pad and your choice of plane groups.

13 Research the life of Evariste Galois. Write a poem or a one act play about him.

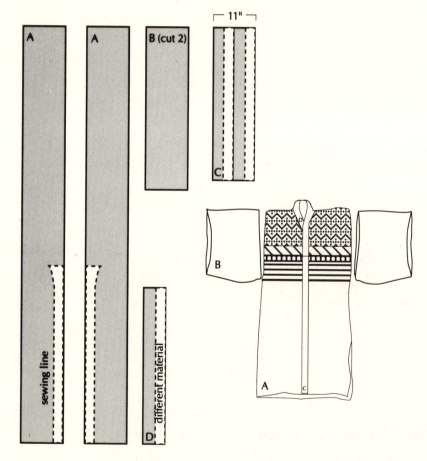

Above:
Fig. 2.21   Kimono Design

76

Further Reading

Finkel, Leslie. *Kaleidoscopic Designs and How to Create Them*. New York: Dover Publications, Inc., 1980.

Gombrich, Ernst. *The Sense of Order*. Ithaca: Cornell University Press, 1979.

Kennedy, Joe and Diane Thomas. *Kaleidoscope Math*. Palo Alto: Creative Publications, Inc., 1978.

Kim, Scott. *Inversions*. Peterborough: Byte Books, 1981.

Schattschneider, Doris. *Visions of Symmetry*. New York: W. H. Freeman and Company, 1990.

Seneschal, Marjorie and George Fleck. *Patterns of Symmetry*. Amherst: University of Massachusetts Press, 1977.

Stevens, Peter. *Handbook of Regular Patterns, An Introduction to Symmetry in Two Dimensions*. Cambridge: MIT Press, 1980.

Washburn, Dorothy K. and Donald W. Crowe. *Symmetries of Culture Theory and Practice of Plane Pattern Analysis*. Seattle: University of Washington Press, 1988.

Weyl, Hermann. *Symmetry*. Princeton: Princeton University Press, 1952.

3 *The Circle in the Plane*

All sacred constructions
represent the Universe in symbolic form.
               Maria-Gabriele Wosien
                       Sacred Dance

In western civilization, the development of geometry has come to be viewed, in part, as a means to understand Nature through the use of the rational and the logical. Other civilizations, especially that of Islam, have perceived geometry to be the way to understand the spiritual and the divine. In a culture where the animal and the human were barred from use in art, geometric pattern became the vehicle of expression for humankind's relationship to the cosmos.

These two positions are not mutually exclusive. Throughout the evolution of civilizations, forms have always held both literal and metaphoric meaning. In accepting both these aspects, the understanding of forms takes on a richness that is lacking when they are experienced solely from one vantage point. Therefore, geometry is simultaneously philosophy, mathematics, and art. If philosophy is the search for essentials, geometry is a description of those essentials, and art the expression of them.

Our species defines itself through symbols. All language making is symbolic. In communication, marks, movements, and sounds become resonators of other ideas. The circle is a prime example of such complexity. It is the archetype for the processes of the Universe. Despite its visual simplicity it is filled with multiplicity. It is a form that has no beginning and no end; it is complete unto itself. It has become the accepted shape of perfection and total symmetry. All things are possible within it, and its meanings are manifold. It denotes neither vertical nor horizontal, but has a strong sense of center with movement toward or away from that point. Its appearance dates back to the primeval period where the sun and the circle became one, suggesting warmth, potency, light, and fertility. It is also connected with the cycles of time yet represents eternity. In cold countries it is full of potential and promise, while in hot countries it is filled with power and destruction.

Lower two sun forms from the works of Jorge Enciso, courtesy of Dover Publications, Inc.

The circle is ubiquitous. It is found on the walls of prehistoric caves, on the embroidery of peasant clothing, on architectural ornamentation, on the shields of warriors, in the illuminated manuscripts of monks, and on the musical instruments of shamans. It is found both large and small. The astronomical observatory of Stonehenge in England, circa 2,000 B.C., is circular

in plan as are the medicine wheels, circa 2,500 B.C., found in the landscape of the Western Plains of the United States and Canada. It is found in the shape of the tipi shelters of North American Indians, the igloo of the Eskimo, and the yurt tents of the nomadic Mongolians of Siberia. It is seen in the stone calendars of the Aztec culture of South America and in the God's eye decorations of Mexico.

> *Geometry – and the use of polygons in setting proportions – is fundamental to all rose windows: every one of them involves extremely careful calculation and precise construction. It operates in different ways in different windows, and in the greatest of them the satisfaction is more than visual – it is intellectual and even cosmic in its implications.*
>
> Painton Cowen
> Rose Windows

Above:
Rose window. North transept of Chartres Cathedral, France. Photograph courtesy of the author-photographer, Painton Cowen.

Gothic architects used the circle in constructing the Rose windows of French cathedrals. These symbols of love and union exemplify the sacred mandala structure; that is, a protective circle with radiating elements that contain a four part symmetry. According to the psychologist C. G. Jung, it is a symbol that expresses the human desire for unity and wholeness. Order is imposed upon the sense experiences and chaos is quieted.

One of the supreme examples of this art form is seen in the north transept of Chartres Cathedral in France, a window dedicated to the Virgin Mary. Number and geometry express a sense of the Divine Order. The number twelve, expression of perfection and the product of three and four, is the key to understanding the system of relationships. Number is translated into a geometric equivalent. As is found in Islamic art, the square within the circle is the generative form. It brings together the sense of the infinite with the finite. Through the use of diminishing squares, there is a connection to the growth of nature's sunflower with its spirals emanating from the center. And these relate to Fibonacci numbers and the Golden Ratio. In Chapter 4 we shall show a visual link to this through a tiling of proportional squares.

In the Italian Renaissance of the 15th and 16th centuries, the tondo, as the circle form is called, was used in painting to suggest the idea of perfect beauty. Twentieth Century artists have also found the circular form meaningful.

Above:
The School of Botticelli. 15th Century. The Nativity. *Isabella Stewart Gardner Museum, Boston.*

Below:
George Tooker. Self Portrait with Shell. *Collection of Mr. and Mrs. Lincoln Kirstein. Photograph courtesy of the artist.*

This page:
Natural forms suggestive of circular shapes.
Top to bottom:
Flower head
Coral
Tree nodes

As we began the exploration of circles, we wondered how humankind first became aware of them. What natural phenomena would suggest the idea? Was it the shape of the sun or the moon; the act of pivoting in space; the reflection of one's eyes in a pool, or the ripples on the surface of water when a stone was tossed in? Whatever physicalities initiated the concept, it had to be abstracted before it could be used; a process of finding similarities and then distilling their essences. Afterwards comes the naming of the abstraction which allows for generalizing. For when a concept is named, it can then be differentiated from other ideas, understood, and then discussed and shared.

Certainly, by the time the wheel was in use, people had perceived the advantage of the circular form. However, there came a time when it was expeditious to impose an outside order upon the form, one that was intellectual rather than natural.

The circle was defined to be the **set** of all points in a plane equidistant from a given point called the **center** of the circle. Geometrically, this figure is stark and neutral. In Western civilization we do not question this neutrality. However, when imbued with the sacred, primarily in Eastern cultures, certain divisions of the circle, together with its **interior**, are of paramount importance. A subdivision into four parts suggests the primary compass points of North, South, East and West; the four elements of Earth, Air, Fire and Water;

the four seasons, and the four human faculties of thinking, feeling, sensing and intuiting. In a more practical vein, other subdivisions were useful. For instance, an analog clock face is divided into twelve hours, each of which contains 60 minutes, which in turn contain 60 seconds. The question is "why?" Math historians suggest that some of these divisions are the result of the use of the sexagesimal (base 60) system of numeration used by the ancient Babylonians, circa 2,000 B.C. This numeration system had its origins in astronomy since it was believed that the year had 360 days, a number easily divisible by 60, 24, and 12. The circle is divided into units called **degrees** and there are 360 of them in every circle (from which the polar grid is developed). This seems to be a more flexible choice than an anatomically based (10 fingers) system, because 360 has many more **factors** than 100. Whereas 360 can easily be broken into halves, thirds, fourths, fifths, sixths, eighths, ninths, tenths, etc., 100 cannot be broken into thirds, sixths, eighths, or ninths without using fractions.

Geometric Properties of Circles

The circle is the very basis of geometric construction, and the compass the essential tool. Whenever an arc is cut with the compass, a portion of a circle is drawn. Points are determined by intersecting arcs which then allow for the existence and measurement of linear entities.

In order to communicate ways in which to use the circle, we need to have a common vocabulary. First, let us consider the way in which the circle subdivides the plane into three distinct regions: the **exterior**, the circle itself, and the **interior** (Fig. 3.1). Note that the center of a circle, by which it is named, is not a part of the circle, but rather belongs to its interior. Fig. 3.2 describes the relationship to the circle of certain lines, line segments, and plane regions.

Below:
Fig. 3.1 A circle subdivides the plane into three distinct regions.

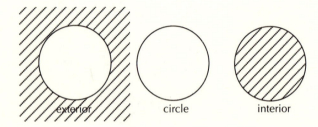

exterior   circle   interior

Right:
Fig. 3.2 Geometric figures in relation to the circle.

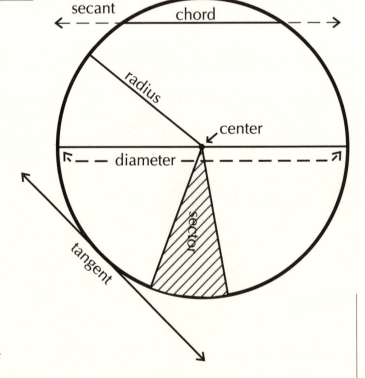

Circumference*: the distance around the circle. This denotes a real number rather than a geometric figure. To find the circumference use the formula: C = πd or C = 2πr. (C = circumference, d = diameter, r = radius, π ≈ 3.14)

Radius*: a line segment from the center to a point on the circle.

Chord*: a line segment which has its endpoints on the circle.

Diameter*: a chord that passes through the center of the circle.

Tangent*: a line that intersects the circle in exactly one point and lies in the same plane.

Secant*: a line that contains a chord of the circle.

Sector*: a region in the interior of the circle bounded by two radii and the circle itself. A sector is called a quadrant if it contains one-fourth of the interior.

To find the ***area*** of the circle use the formula: A = πr² (A = area, r = radius, π ≈ 3.14).

The following three constructions each use the property that the perpendicular bisector of any chord passes through the center of the circle. Therefore, the perpendicular bisectors of any two chords will intersect at the center.

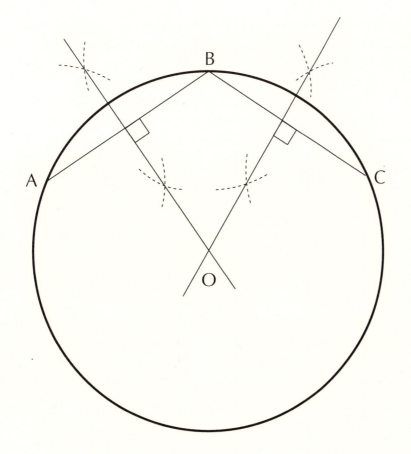

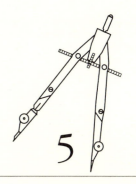

Find the Center of a Circle

Given a circle.

1. Draw two chords anywhere on the circle ($\overline{AB}$ and $\overline{BC}$ in the diagram).

2. Construct the perpendicular bisectors of each of the chords. *(See Construction 3, Book 1, Universal Patterns.)*

3. Label the point of intersection O.

Now O is the center of the circle.

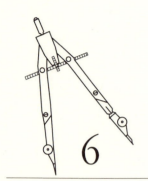

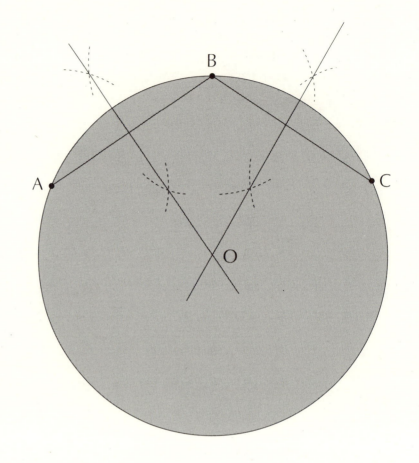

6

Construct a Circle through Any Three Noncollinear Points

Given points A, B and C not on the same line.

1. Draw $\overline{AB}$ and $\overline{BC}$.

2. Construct the perpendicular bisectors of $\overline{AB}$ and $\overline{BC}$, respectively, *(See Construction 3, Book 1, Universal Patterns.)* and extend them so they intersect in a point.

3. Label the point of intersection O.

4. Place the metal tip of the compass on O and the pencil tip on A (or B, or C) and draw a circle.

Now circle O passes through each of the points A, B and C.

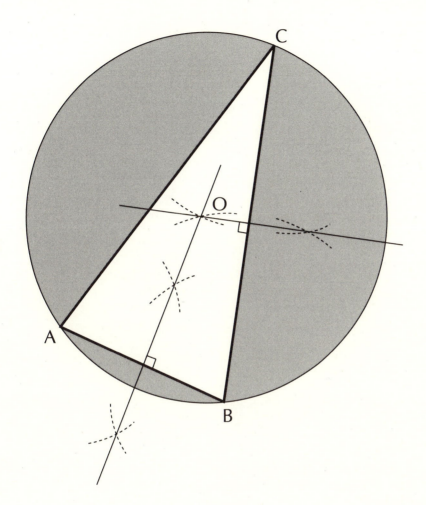

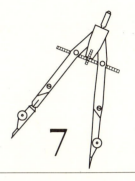

7

Circumscribe a Circle about a Triangle

Given △ ABC.

1. Construct the perpendicular bisectors of any two of the sides of the triangle, and extend them to meet in a point.

2. Label the point of intersection O.

3. Place the metal tip of the compass on O and the pencil tip on A (or B or C) and draw a circle.

Now circle O is circumscribed about △ ABC.

Also, given any triangle, a circle may be inscribed within it. This seems logical since an inscribed circle is tangent to each of the three sides. This construction shows how to find those points of tangency.

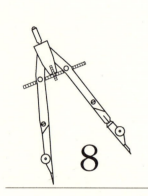

Inscribe a Circle within a Triangle

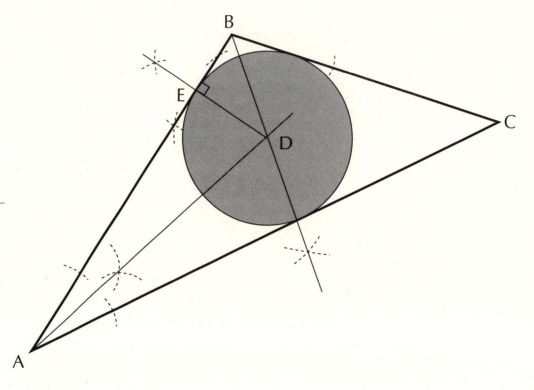

Given △ ABC.

1. Bisect any two angles of △ ABC, in this case angles A and B. *(See Construction 4, Book 1, Universal Patterns.)*

2. Label the point of intersection of the angle bisectors D.

3. Construct a perpendicular from D to any one of the three sides of △ ABC, in this case $\overline{AB}$. *(See Con-*

struction 6, Book 1, Universal Patterns.)

4. Label the point of intersection E.

5. $\overline{DE}$ will be a radius of the inscribed circle, so place the metal tip on D and the pencil tip on E and draw the circle.

Now circle D is inscribed in △ ABC.

Armatures of the Circle

When people possess little information, they tend to produce works mechanically. However, personality, skill and idea coalesce when one is empowered with knowledge. As with the rectangle discussed in Chapter 4, Book I, *Universal Patterns*, we can also consider various **armatures** of the circle in which certain subdivisions give harmonic relationships, and imbue the circle with a sense of spirit.

Because working with circles can be technically tricky, we offer some helpful studio tips before embarking on our discussion of the armature.

1. Sharpen the pencil tip in the compass to a long taper in order to help insure accuracy. This can be done with a piece of fine grade sandpaper, a sanding block or an emery board.

2. If you do not have a manufactured compass, there are simple ways to improvise one:

Use a strip of illustration board with a push pin or thumbtack in one end, to mark the center, and a hole at the other end for the pencil. The distance between the two determines the radius of the circle.

Old wooden rulers, clear plastic quilting rulers, or strips of wood or heavy cardboard can be adapted in a similar manner. Small blocks of Fomecore or illustration board can be attached to the underside of the pencil end to avoid smudging.

3. To prevent damaging the art surface when using a compass, adhere, with a drop of rubber cement, a small square of illustration board at the place where the metal tip would pierce the surface. Remove it after the circle is drawn and clean away the rubber cement.

4. There are compasses which can be equipped with a cutting blade for use on heavier weight stock where scissors are not practical.

5. Work small to large when drawing concentric circles in order to obtain greater accuracy.

6. Some compasses will allow for the insertion of technical pens, markers, brushes, etc. If yours does not, try attaching the tool to the pencil side with a rubber band. If using a brush, be sure to set the radius before dipping it in the paint or ink.

Below:
Fig. 3.3   Inscribing a circle within a square.

The simplest way to construct an armature for the circle is to use straight line segments. Since these are determined by their endpoints, it is helpful to have an easy method for locating appropriate points on the circle. One way to do this is to inscribe the circle within a square.

1. The intersection of the diagonals locates the center of the circle.

2. A segment from the center to the midpoint of any side gives the radius.

3. Set the compass by the radius and let her rip!

1.

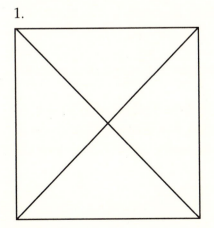

2.

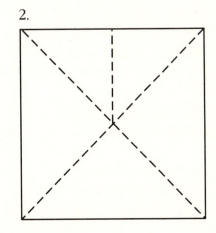

3.

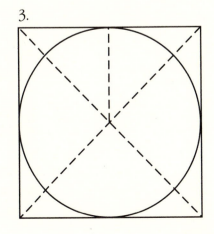

This construction shows how to circumscribe a square if the size of the circle is already determined.

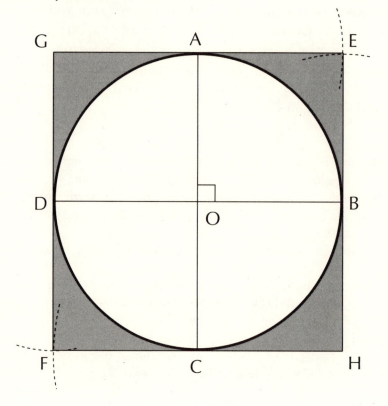

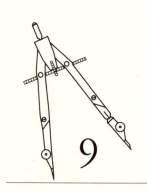

9

Circumscribe a Square about a Circle

Given a circle.

---

1. Locate the center *(Construction 5)* and construct perpendicular diameters.

2. Using OA for the compass setting, place the metal tip on A, B, C and D consecutively and cut two pairs of intersecting arcs in the exterior of the circle to locate opposite vertices of the square. (Use A and B for one pair, C and D for the other.) Label points of intersection E and F.

3. Draw $\overrightarrow{EA}$, $\overrightarrow{EB}$, $\overrightarrow{FD}$ and $\overrightarrow{FC}$ and extend them to intersect at the other two vertices of the square, G and H.

Now square FHEG is circumscribed about the circle.

Studio Tip: There are templates available which are marked to show endpoints of perpendicular diameters in circles and midpoints of sides of squares. When the size of the two figures coincide, the templates are much easier to use than compass and straight-edge.

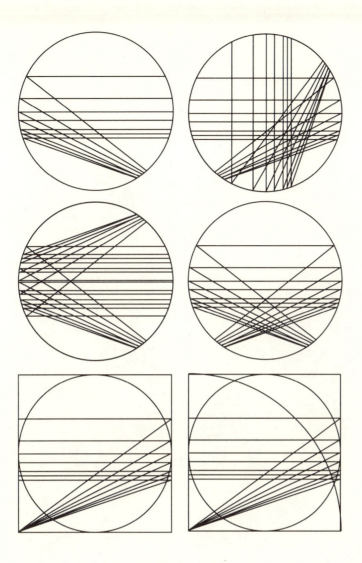

This page:
Fig. 3.4 The sides of the square can be subdivided harmonically using Ø-Proportions, Fibonacci Numbers, etc., or the square itself may be subdivided using an harmonic armature. (Refer to Book I, Universal Patterns for detailed information.) When points are connected on opposite or adjacent sides of the square, the line segments, in most cases, will intersect the circle and it will inherit harmony from its parent square. The examples illustrate some of the ways that this can be done.

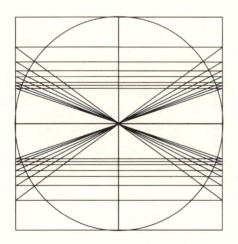

This page:
Eric Kulin. Student Work.
Pieced, photocopied, hand
drawn image.

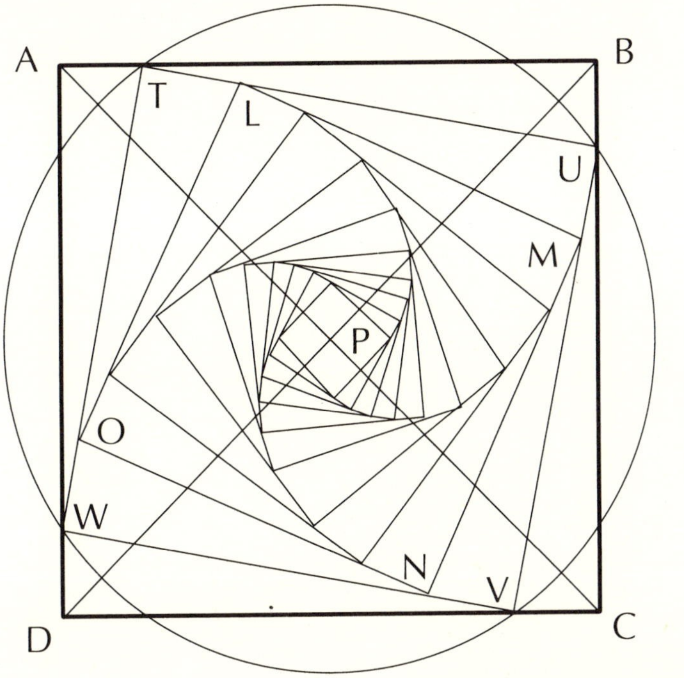

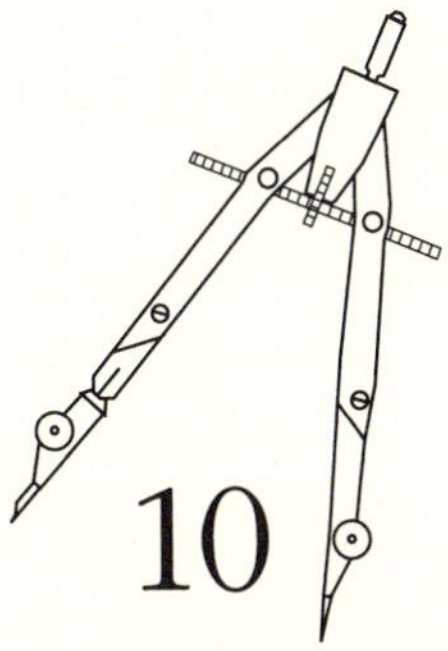

10

Construct a Straight Line Spiral in a Circle Given a Square

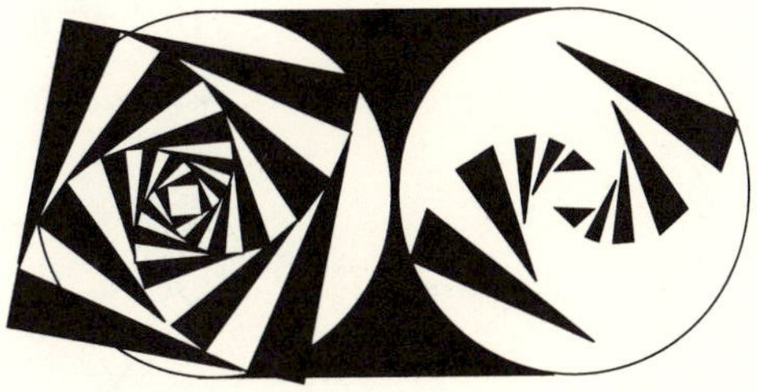

Given square ABCD.

1. Locate point T one-sixth, an arbitrary choice, of the distance from A to B ***(See Construction 8, Book 1, Universal Patterns).***

2. Draw the diagonals of ABCD and label their point of intersection P.

3. Open the compass to measure PT, and with the metal tip on P, draw a circle.

4. Label every other point of intersection of the circle and the square (U, V and W, respectively) travelling in a clockwise direction from T.

5. Join T, U, V and W with consecutive line segments to form a second square.

6. Locate points L, M, N and O one-sixth of the distance from T to U, U to V, V to W and W to T, respectively.

7. Connect the points to form square LMNO.

8. Repeat the process until the last square constructed is the desired size.

Now the circle contains a straight line spiral.

Shading can vary the effect of the spiral in the circle.

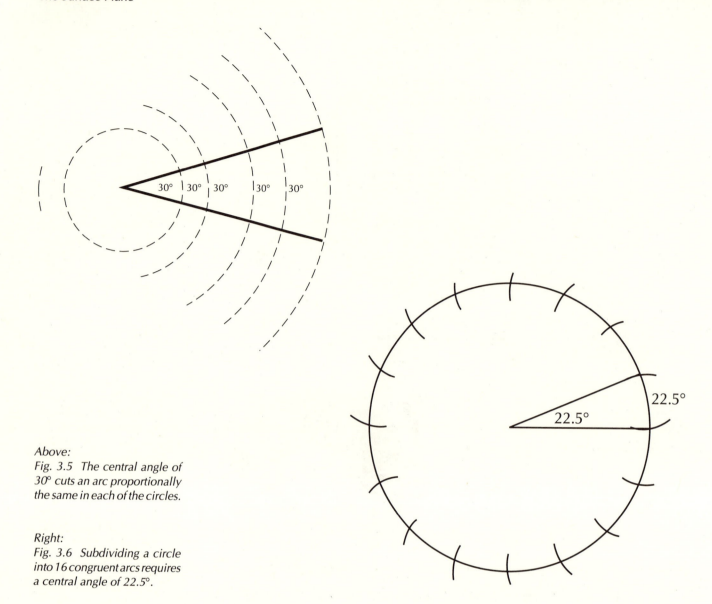

Above:
Fig. 3.5 The central angle of 30° cuts an arc proportionally the same in each of the circles.

Right:
Fig. 3.6 Subdividing a circle into 16 congruent arcs requires a central angle of 22.5°.

Another way to subdivide the circle is to break it into congruent arcs. With the exception of certain special numbers such as three, four, six or eight, this is easier to do with a protractor than it is to use compass and straightedge, and many times it is the only way that it can be done. The number of degrees in the measure of an arc of a circle is the same as the number of degrees in the measure of the **central angle** that determines the arc. We can use the protractor to measure and draw the central angle that gives the appropriate arc length. Note that the size of the circle does not matter (Fig. 3.5).

Since we know that every circle contains 360°, we need only to divide 360° by the number of congruent arcs required to find the measure of the central angles. For example, to divide a circle into 16 congruent arcs, we divide 360° by 16 to get 22.5°. We draw a central angle of 22.5° and then cut the circle consecutively with the corresponding arc length using a compass to obtain the 16 congruent arcs (Fig. 3.6). Greater accuracy can be gotten by working in both directions from the sides of the angle.

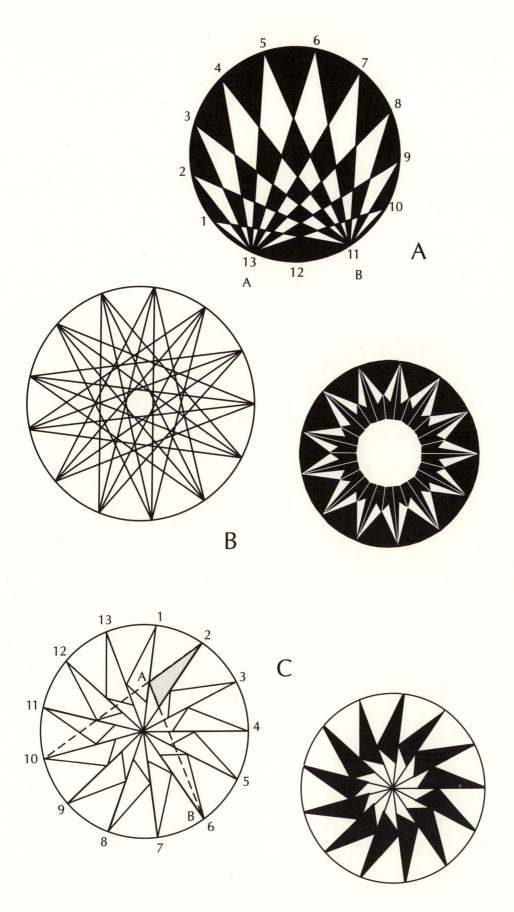

This page:
Fig. 3.7 These figures illustrate how to begin to tap the potential of using Fibonacci numbers for subdivisions. In each case the circle is divided into thirteen congruent arcs using a central angle of 27.7° (360° divided by 13). Points mentioned refer to endpoints of the arcs and there are thirteen of those as well.

Top to bottom:
A. Two points are chosen such that one lies between them. Line segments are drawn from each of the two points to every other point on the circle, with the exception of the other chosen point and the one that lies between them. Regions are shaded alternately.

B. Connect each point to four points on the opposite side of the circle. In the finished design, the line segments in the center are eliminated.

C. Radii are drawn to each of the thirteen points. Then, from each point, a segment is drawn to the fifth point from it. The dotted portion is omitted. The third side of the shaded triangle is gotten by connecting point A to point B, which is the endpoint of the fourth radius from A. Again, the dotted part is omitted. This process draws upon the concept of the "divergency constant" found in the examination of phyllotaxis (see Chapter 5, Book 1, Universal Patterns).

In addition to their use for non-illusionistic design, these armatures can provide the skeletal structures for representational work in painting, drawing and collage. Charles Bouleau, in his book, *The Painter's Secret Geometry,* utilizes the concept of straight line armatures in circles in his analyses of Italian Renaissance tondo figurative paintings. We refer you to this excellent source for further information.

Another way to subdivide a circle is by using others within it, the simplest being a series of **concentric circles.** Fig. 3.8 shows a regular increase in the length of the radii for successive circles, and the use of a central angle of 10° to subdivide the circles into 36 congruent arcs. Notice how by adding diagonals within the units, many options are opened to the designer, among which are the spirals shown.

Other satisfying effects can be achieved if attention is paid to the ratio of successive segments along a radius. When these segments are increased in Fibonacci progression a different harmonic figure results.

Below:
Fig. 3.8 Concentric circles and simple repetition provide a framework for design.

Above:
Mary Rowe. Knitted tam showing circular armature. Knitwear Designs, Stevens Point, WI. Photograph courtesy of the artist.

Below:
Fig. 3.9 Displacement of the center along a diameter shifts the symmetry.

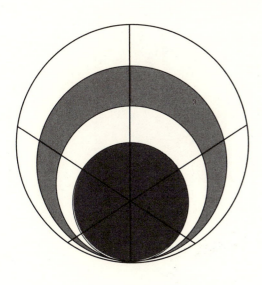

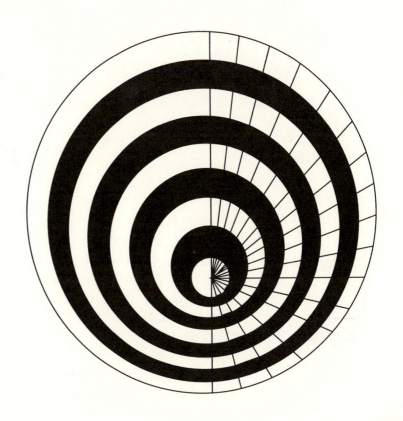

A sense of asymmetry can be introduced into this most perfectly symmetric figure by the displacement of other centers along a diameter. In actuality, the symmetry is not lost but shifts from pointwise to bilateral. Two of the possibilities are given in Fig. 3.9.

Above:
Jackie Peters. Student work.
Ink on paper.

Below:
Fig. 3.10 Chinese Yin/Yang symbol.

Let us now consider armatures which utilize tangent rather than nested circles. These can be examined systematically beginning with what we have named the **internal dyad**. It is the armature consisting of two tangent circles set within another circle. Construction 11 gives the details.

One of the most powerful symbols to use this armature is the ancient Chinese Yin/Yang. Referring to cosmic forces, it represents the union of opposites: male-female; positive-negative; life-death; etc. It is gotten by eliminating portions of the internal circles and shading appropriately as shown in Fig. 3.10.

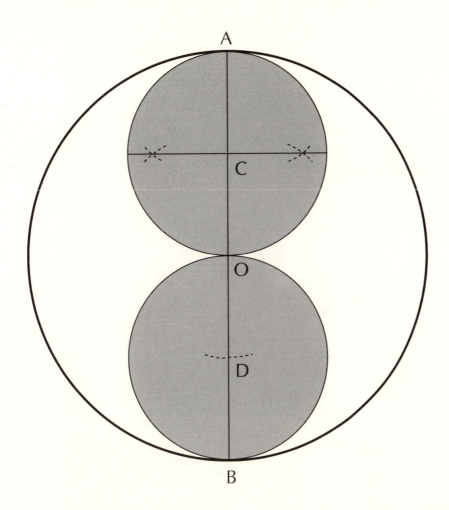

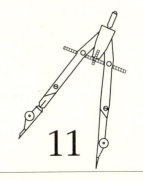

11

Construct an Internal Dyad

> Given circle O.

1. Draw a diameter and label its endpoints A and B.

2. Find the midpoint of $\overline{AO}$ and label it C.

3. Place the metal tip on O and cut an arc on $\overline{OB}$ with radius OC.

4. Label the point of intersection D.

5. Without changing the compass setting, draw circles C and D.

Now circle O contains an internal dyad.

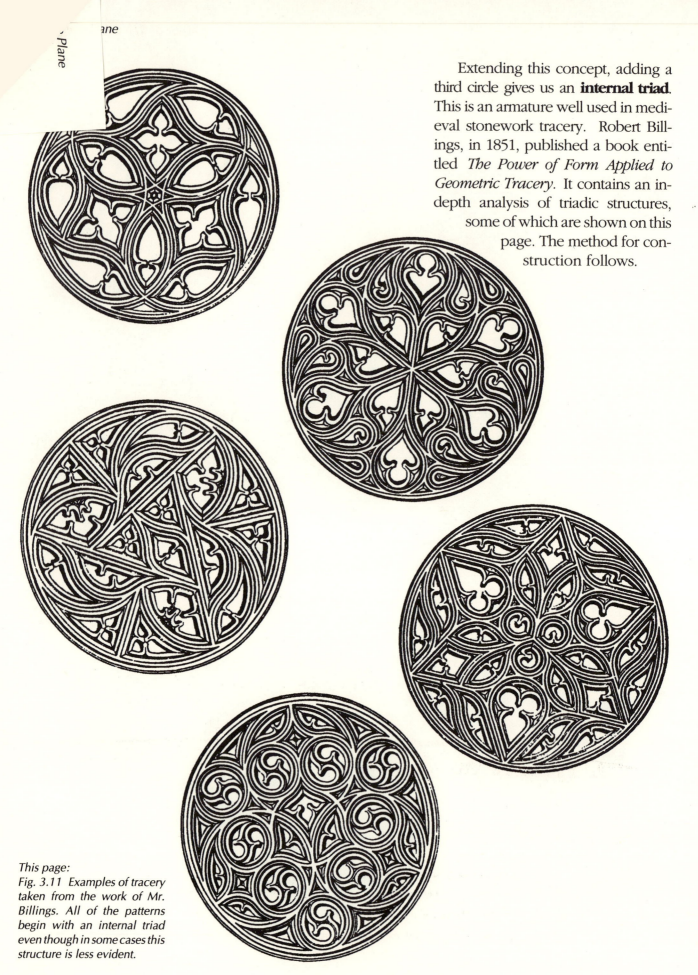

Extending this concept, adding a third circle gives us an **internal triad**. This is an armature well used in medieval stonework tracery. Robert Billings, in 1851, published a book entitled *The Power of Form Applied to Geometric Tracery*. It contains an in-depth analysis of triadic structures, some of which are shown on this page. The method for construction follows.

This page:
Fig. 3.11 Examples of tracery taken from the work of Mr. Billings. All of the patterns begin with an internal triad even though in some cases this structure is less evident.

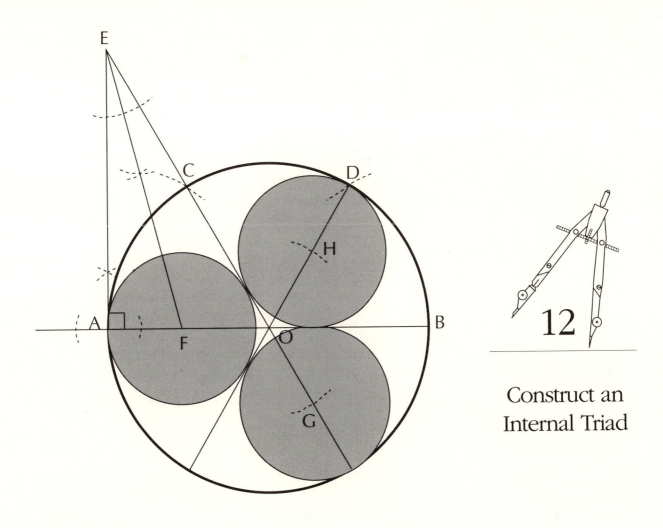

12

Construct an Internal Triad

Given circle O.

1. Draw a diameter, label its endpoints A and B, and extend it through A.

2. With the compass open to the radius of circle O, place the metal tip on A and mark point C on the circle.

3. Repeat at B to get point D.

4. Draw $\overrightarrow{CO}$ and $\overrightarrow{DO}$ and extend them to intersect the opposite side of the circle.

5. Construct a perpendicular to $\overleftrightarrow{AB}$ at A, draw $\overrightarrow{OC}$, and extend both until they intersect at a point (E).

6. Bisect $\angle$ AEC and extend the bisector to intersect $\overline{AB}$ at a point, F.

7. Open the compass to measure OF, place the metal tip on O and cut points G and H as shown. F, G and H will be the centers of the three circles of the triad.

8. Using FA as the measure of the radius, draw circles F, G and H.

Now circle O contains an internal triad.

Internal Tetrad

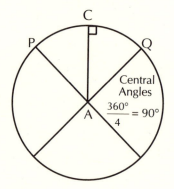

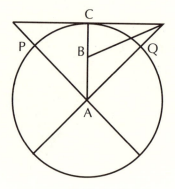

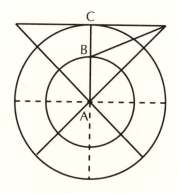

It seems reasonable to extend this idea to other multicircular armatures, and, in fact, it is not too complicated to do so. Remembering that a circle can be inscribed in *any* triangle (Construction 8), the problem becomes one of determining the appropriate triangle so that the circle may be repeated the desired number of times. In Construction 12, Δ OEA is actually one-half of the triangle in which circle F is drawn. Since all three angle bisectors of a triangle meet at a point, only two of them are necessary to determine that point.

In this process, we are actually creating point groups in which the circle functions as the motif, and each circle is tangent to its neighbors. The chart diagrams this procedure for other multicircular armatures. In each case the essential triangle has been shaded.

You might note that one side of this triangle is a side of a regular polygon that circumscribes the circle. The number of sides corresponds to the number of circles in the armature. Therefore, if you have a captive regular polygon, you can construct internal circular armatures by drawing the diagonals through the center. Then follow the last three steps given. This process may be continued for as many internal circles as desired.

Internal Pentad

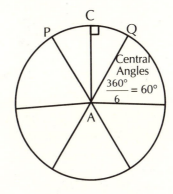

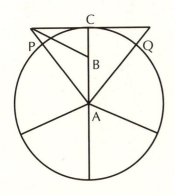

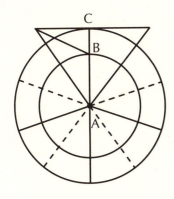

Internal Hexad

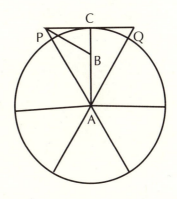

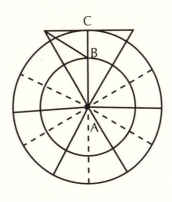

The number of desired circles in the armature determines the number of sectors required for the initial subdivision. The central angle PAQ is bisected and a perpendicular is drawn to the bisector, tangent to the circle at C.

The triangle is drawn by extending the radii forming the sector to intersect the tangent line. The center of the inscribed circle is determined (Construction 8). BC is the radius.

The centers of the remaining circles in the armature are also centered on the bisectors of the central angles of their respective sectors.

AB is the distance of the centers from the center of the original circle and BC determines the radius.

This page:
*Fig. 3.12 Analysis of the struc-
ture of stone tracery taken from
the work of Mr. Billings.*

1. Construct the internal triad.

*2. Within each circle of the
triad, construct an internal
hexad.*

*3. Construct a line through C,
parallel to $\overline{AB}$ (Construction 7,
Book 1, Universal Patterns).
Locate D on the parallel just
constructed such that AC =
BD. Circles C and D are congruent.*

*4. Repeat for the remaining
five spaces between the circles of the triad using centers
of the appropriate circles in
the internal hexads to construct
parallels.*

5. Embellish.

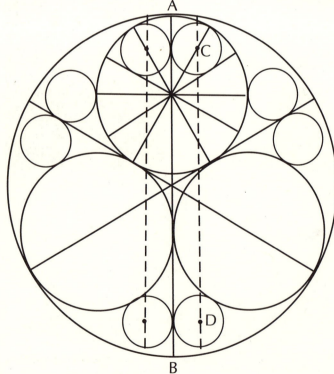

Left:
Fig. 3.13   Other kinds of complexity, which add an aesthetic component to pure structure, can be achieved using the concept of interlacing, Fibonacci or other harmonic subdivisions, or shifting and overlapping congruent circles.

Center:
Circular armatures as seen in the art of gravestones.

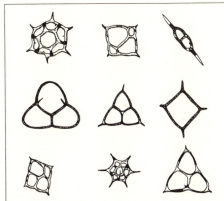

Left:
Fig 3.14   The notion of a multicircular armature is suggested by a group of plankton known as Silicoflagellata. They are a single-celled species covered by a simple skeleton-like structure which forms a protective covering.

107

Close Packing of Circles

At times there is a need to go beyond the limits of oneness, beyond internal divisions. When a problem can be solved by the addition of a small number of units, a linear arrangement, like a chain, may provide an answer. However, when a planar surface is to be covered with large numbers of units, the chain is not efficient and there is a need for units to cluster in an organized fashion.

In living organisms, the system generally requires more than a single cell. As more material is needed and more complexity and specialization required, the cell splits rather than becoming larger. Clusters of cells can grow so that the organism can reach the prescribed size and shape. Although the nature of specialization determines its shape, the size of the basic cellular building block remains fairly constant.

Abstracting the essence of this process, we will look at what happens when circles are packed closely in the plane. There are only two basic arrangements possible. In Fig. 3.15, the diagram on the left shows the pattern that would result from filling the squares in a grid with circles, whereas the one on the right illustrates an isometric grouping.

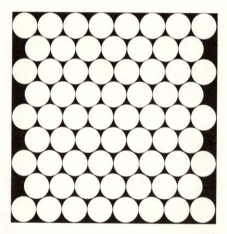

Above:
Fig. 3.15 Square and isometric packing of circles.
Below:
Fig. 3.16

Let us now consider the first arrangement. Could we cover more of the surface area with smaller circles placed in this manner? To find an answer a little mathematics is helpful.

Suppose we have a square with side of length one (1) and a circle inscribed within it (Fig. 3.16).

The area of the square is 1^2 or 1, since $A = s^2$.

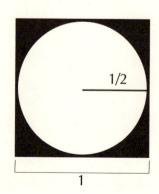

The area of the circle is πr^2 where the radius, r, is 1/2 (half the length of the side of the square).

$$A = \pi\left(\frac{1}{2}\right)^2 = \pi\left(\frac{1}{4}\right) = \frac{\pi}{4}.$$

Therefore, the ratio of the area of the circle to the area of the square is $\pi/4$ to 1 or $\pi/4$.

Fig. 3.17

As with the pebbles of Pythagoras, circles can only be placed within this square in "square" arrangements. Therefore, the next collection would contain four circles (Fig. 3.17).

Now the radius of each circle is 1/4, so the area of each is:

$$A = \pi\left(\frac{1}{4}\right)^2 = \pi\left(\frac{1}{16}\right) = \frac{\pi}{16}.$$

There are four circles so their total area is $4(\pi/16)$ or $\pi/4$. Again the ratio of the area of the circles to the area of the square is $\pi/4$.

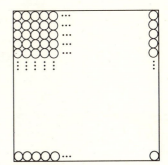

Fig. 3.18

The next perfect square is nine. If we use nine circles, each row contains three (Fig. 3.18). The diameter of each is 1/3 and the radius is half of that or 1/6. The area of each circle is:

$$A = \pi\left(\frac{1}{6}\right)^2 = \pi\left(\frac{1}{36}\right) = \frac{\pi}{36}.$$

and the total area of the circles is $9(\pi/36)$ or $\pi/4$. Once again, the ratio of the area of the circles to the area of the square is $\pi/4$.

At this point we could do a little inductive reasoning and make an educated guess that even if we decrease the size of the circles again and again, we will never be able to cover more of the interior of the square with them. It appears as if the ratio of the area of the circles to the area of the square will remain constant at $\pi/4$. We can take all the guesswork away by using the process of mathematical proof.

Proof:

Assume we have a square with a side of length one (1) and that we cover its interior with *n* tangent circles. We know that *n* must be a perfect square (Fig. 3.19). Then every row contains *x* circles and $x^2 = n$. The diameter of each circle must be 1/x and the radius must be half of that, or 1/2x.

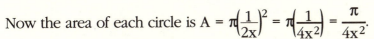

Fig. 3.19

Now the area of each circle is $A = \pi\left(\frac{1}{2x}\right)^2 = \pi\left(\frac{1}{4x^2}\right) = \frac{\pi}{4x^2}.$

But this equals $\pi/4n$ since $x^2=n$, and since there are *n* circles all together, the total area of the circles is $n(\pi/4n)$ or $\pi/4$. And the ratio of the area of the circles to the area of the square is, indeed, $\pi/4$. Now we are certain that our guess was correct.

And that is the great advantage of proof!

Center:
Fig. 3.20 Transformation from closely packed circles to hexagons.

Bottom:
A pan of circular snickerdoodle cookies become polygonal while baking when they are packed too closely.

Counting the number of circles in each square in Fig. 3.15, we can see that the isometric one is more efficient, having 68 as opposed to 64 circles. Despite the fact that of all plane figures, the circle encloses the greatest area for a given perimeter, circle packing is not the most efficient way to cover the plane. When circles are organized isometrically and then more tightly packed, the circular form is compressed and, with sufficient pressure, takes the form of a hexagon. The transformation is pictured in Fig 3.20.

The hexagon provides the simplest and most economical way to cover the entire plane with units of uniform size and shape. One of Nature's most familiar examples is the honeycomb where the tubular cells give maximum storage space for minimum use of wax, and a cross-section shows the hexagonal structure.

Closest packing, a universal pattern, occurs independent of scale. Materials change, but the pattern remains constant. As with other natural phenomena, mathematical perfection is affected by such factors as sequence of growth and boundaries. Any natural surface that is closely packed may contain different polygons, but the most often occurring one will be the hexagon.

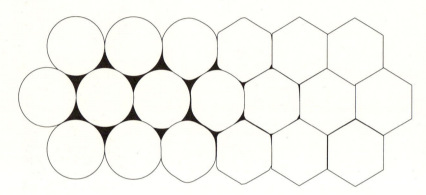

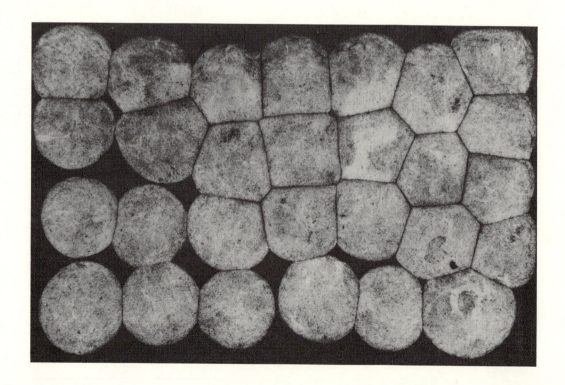

The Surfa

Richard Newman. Stacked Polygons. *Ink on paper.*

Circles and Polygons

The circle may be considered the parent of all regular polygons. We have already seen how a circle may be subdivided into any given number of congruent arcs. Certainly, if the end-points of the arcs are joined in succession by line segments, a regular polygon results. We shall now look at a detailed compass and straight-edge method for generating some of the regular polygons with the same length sides. In the next chapter we shall see how useful such polygons are.

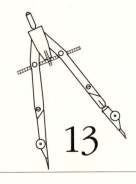

13

Construct a Family of Regular Polygons Having Sides of the Same Length

Given $\overline{AB}$.

1. Construct two congruent intersecting circles A and B with radius $\overline{AB}$. (Extend $\overleftrightarrow{AB}$ for future use).

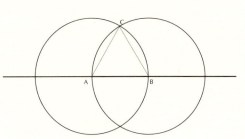

2. Complete the equilateral triangle by joining A and B, respectively, to C, one of the points of intersection of the circles.

3. Draw the perpendicular bisector of $\overline{AB}$ by joining C and the other point of intersection of the circles, and extend it well beyond the two circles.

4. Erect perpendiculars to $\overleftrightarrow{AB}$ at A and B and extend them parallel to the line drawn in Step 3. (The extensions are shown as dotted lines in the diagram.) Complete the square by joining D and E, the points of intersection of the circles and the perpendiculars just constructed.

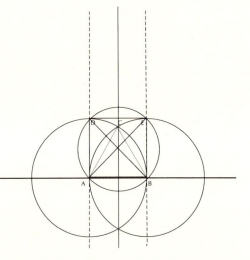

5. The point of intersection, F, of the diagonals of the square ($\overline{DB}$ and $\overline{AE}$) is the center of the circle circumscribed about the square. By putting the metal tip on F and the pencil tip on any of the vertices of the square, the circle can be drawn.

6. To construct the pentagon:
    a. Place the metal tip on the midpoint of $\overline{AB}$, the pencil tip on D, and cut a semicircle with endpoints Y and Z on $\overleftrightarrow{AB}$.

    b. Place the metal tip on A, the pencil tip on Z, and cut an arc that intersects $\overrightarrow{FC}$ at W. Repeat with pencil

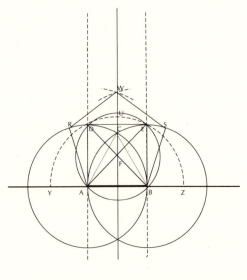

tip on Y and metal tip on B. W should lie on both arcs just constructed and on $\overrightarrow{FC}$. Check your accuracy at this point. These two arcs intersect circle A at R and circle B at S. Draw $\overline{AR}$, $\overline{RW}$, $\overline{WS}$ and $\overline{SB}$ to complete the pentagon.

7. We now have the basis for constructing the circles containing the hexagon, the octagon, the decagon and the dodecagon. Each circle is centered on $\overrightarrow{FC}$ and its radius can be gotten by placing the metal tip on the center and the pencil tip on either A or B.

    a. C is the center of the circle circumscribed about the hexagon.
    b. U is the center of the octagon.
    c. W is the center of the decagon.
    d. V is the center of the dodecagon.

8. Once the circle is drawn, the sides (of length AB) can be marked off in succession. Each of the above mentioned polygons will have a side parallel to $\overline{AB}$ with its endpoints on $\overrightarrow{AD}$ and $\overrightarrow{BE}$. Therefore, when marking off the sides, greater accuracy can be achieved by working in both directions around the circle from A and B and the endpoints of the side parallel to $\overline{AB}$.

Now the family of polygons with same length sides is complete.

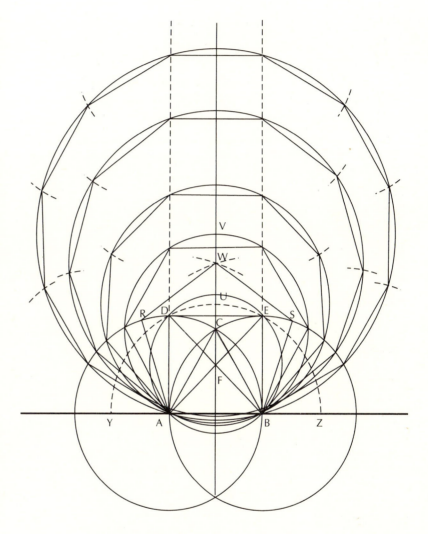

Note: The point of intersection of $\overrightarrow{FC}$ and the circle circumscribed about any polygon is the center for the circle circumscribed about the polygon with twice as many sides.

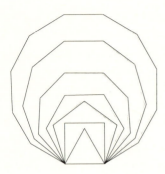

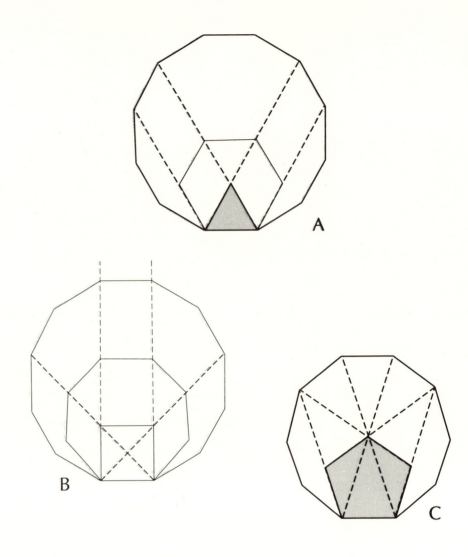

This page:
Fig. 3.21 Nested regular polygons. The diagrams show how extensions of sides and diagonals of the smaller polygons intersect key points on the larger ones.

A. triangle, hexagon and dodecagon.

B. square, octagon and dodecagon.

C. pentagon and decagon.

D. the relationship between all the above mentioned polygons.

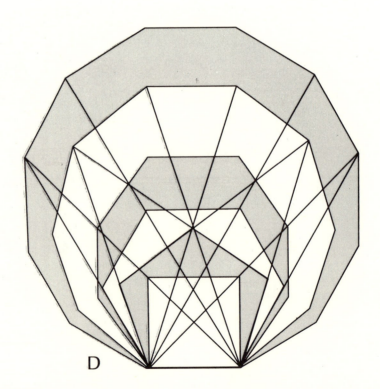

Let us re-examine the two intersecting circles that form the basis
for Construction 13. The figure is known as the Vesica
Pisces, or vessel of the fish,
which is associated in the
Christian religion with
Christ who lived in the
age of Pisces. It is also
an important symbol in
Eastern philosophies as
an expression of the inte-
gration of polarities. Mathematically it resembles a
Venn diagram for two sets
whose common elements (the
intersection of the two sets) fall within
the central portion (see Fig. 3.22).

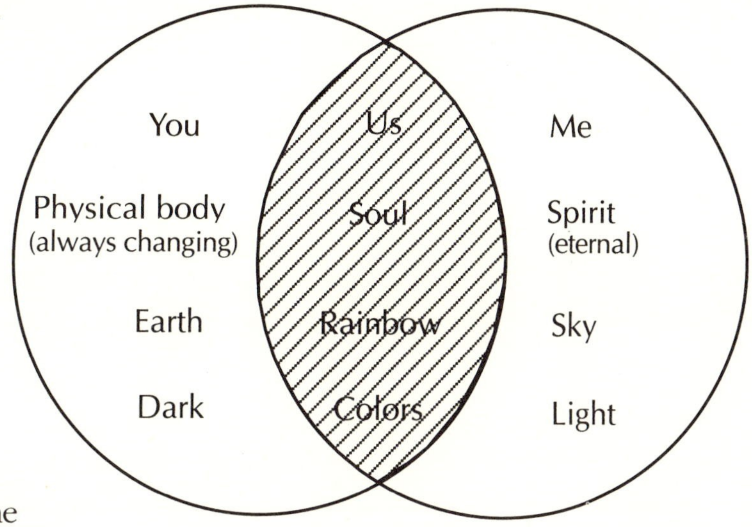

Above:
Fig. 3.22 The Vesica Pisces

Below:
Fig. 3.23 The Vesica Pisces also has some interesting connections to some previously explored ideas.
Open the compass to measure the diameter of either circle. Place the metal tip on A and draw an arc that intersects the other circle in two points (D and E).

Δ DCE is the familiar one characterized by a cross section of the Great Pyramid at Giza, Egypt (see Chapter 3, Book I, Universal Patterns).
By the property of similarity, this same triangle appears as CFG.
CAG and CBE are Triangles of Price.

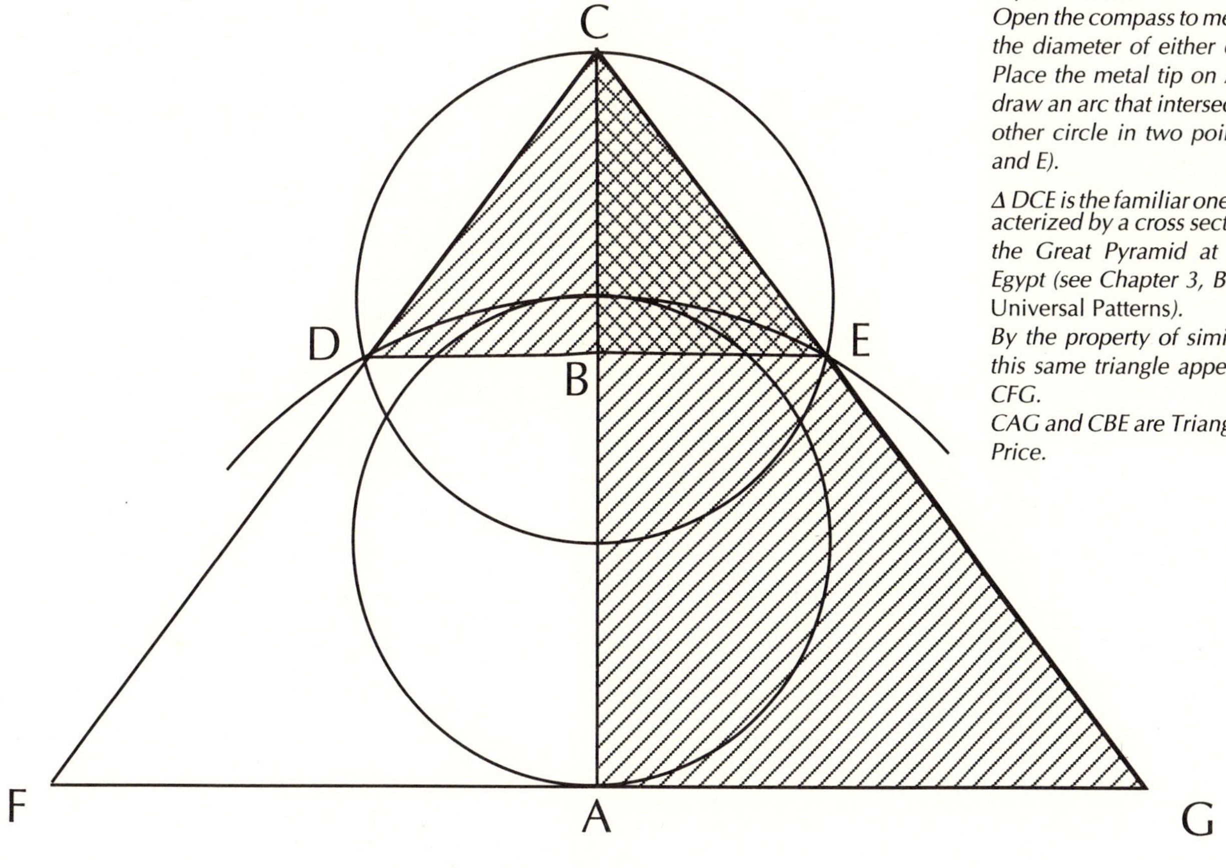

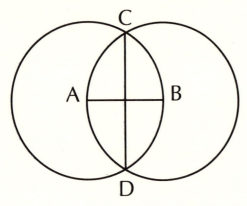

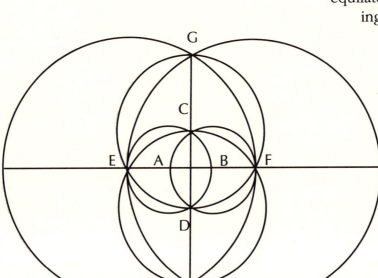

If lines are drawn through the centers of the circles and through their points of intersection ($\overleftrightarrow{AB}$ and $\overleftrightarrow{CD}$, respectively), then the ratio of CD to AB is $\sqrt{3}$ to 1 (Fig. 3.24). If the line segment through the points of intersection becomes one connecting the centers of a larger pair of circles, and the process repeated as often as desired, the $\sqrt{3}$ ratio is preserved.

Likewise, the sides of the pairs of equilateral triangles, formed by joining the endpoints of each pair of segments in succession are in $\sqrt{3}$ proportion (Fig. 3.25). Since it is the case that the ratio of the areas of two similar triangles is equal to the square of the ratio of the corresponding sides, we have that the areas of the successively smaller triangles are in $\sqrt{3}^2$ proportion. That is, for any consecutive pair of triangles, the larger has an area three times that of the smaller (Fig 3.26).

Top to bottom:

Fig. 3.24
$$\frac{CD}{AB} = \frac{\sqrt{3}}{1}.$$

Fig. 3.25
$$\frac{GH}{EF} = \frac{EF}{CD} = \frac{CD}{AB} = \frac{\sqrt{3}}{1}.$$

Fig. 3.26
$$\frac{area\,\Delta\,IGH}{area\,\Delta\,GEF} = \frac{area\,\Delta\,GEF}{area\,\Delta\,EDC}$$
$$= \frac{area\,\Delta\,EDC}{area\,\Delta\,CAB} = \frac{3}{1}.$$

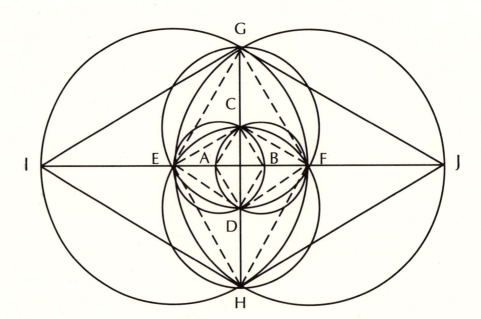

Should one need to fit a regular polygon within a given space, the easiest way to do it is to use a circle that most closely fills the space, and inscribe the polygon within it. This differs from the method described in Construction 13 in that it gives control over the final overall size of the polygon as opposed to controlling the length of its sides.

The simplest one to obtain is the hexagon, and it is frequently, although sometimes erroneously, assumed that everyone knows how to do it. Just because it is simple does not mean that it is self-evident.

The regular hexagon has the property that the length of its side is the same as the radius of the circumscribed circle. Therefore, the sides can be marked off without changing the compass setting once the circle is drawn. If alternate points marked on the circle are joined, an equilateral triangle is obtained (Fig. 3.27).

If the endpoints of a pair of perpendicular diameters are joined in succession a square emerges. Further, if the perpendicular bisector of a side of the square is constructed, a point on the circle is obtained which can be used to determine the length of a side of the regular octagon (Fig. 3.28).

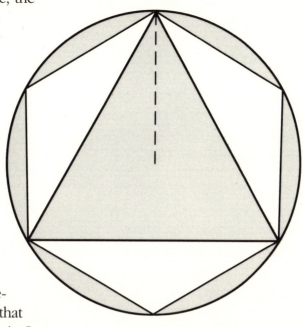

Above:
Fig. 3.27 Hexagon and triangle inscribed in a circle.

Below:
Fig. 3.28 Square and octagon inscribed in a circle.

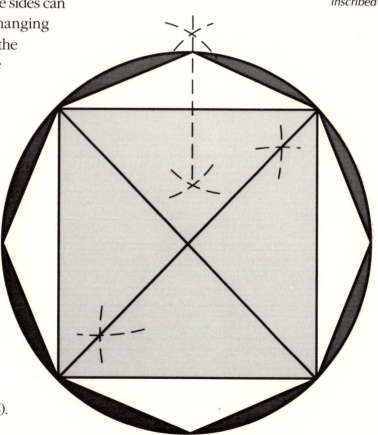

117

In Book I, *Universal Patterns*, Construction 11 gives the steps for constructing the pentagon, a much more complicated operation than those just discussed. As with the square and octagon, however, the decagon may be obtained by bisecting a side of the pentagon. Given this information a great many polygons can be constructed, since it is always possible to double the number of sides using perpendiculor bisectors. The new polygon inherits the properties, harmonic and dynamic, of the parent. Golden properties are especially evident in the progeny of the pentagon which, as you recall, together with its pentagram has many pairs of segments whose ratios are Ø.

In the decagon, the ratio of the radius to the side is Ø. For a mathematician this is not a statement to be taken at face value without asking "why?" In fact, for all human beings it is important to question and to verify so as not to be at the mercy of someone else's statement about what is true. In mathematics, verification comes from the vehicle of proof.

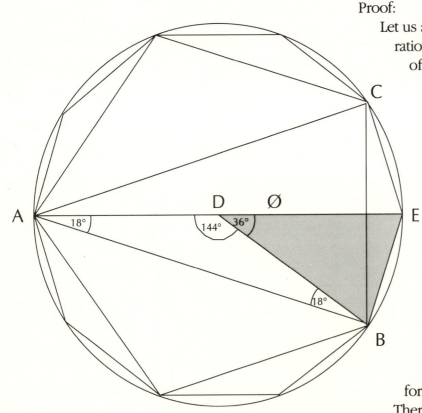

Fig. 3.29 The ratio of the radius of the regular decagon to the side is Ø.

Proof:

Let us answer the question, "Why is the ratio of the radius of a circle to a side of an inscribed regular decagon Ø?" Assuming that the answer to this question is not "Who cares?" let us proceed.

In Fig. 3.29 we wish to show that DE = Ø. We can do that if we can show that Δ DBE is a Golden Triangle. We know:

Δ ABC is Golden since it comes from the pentagram.

Therefore m∠ BAD = 18° (half the measure of the vertex angle of 36° in a Golden Triangle).

Δ DAB is isosceles since it is formed by two radii of the circle.

Therefore, ∠ DBA must also measure 18°. Since the angle sum of a triangle is 180°, m∠ ADB = 144°.

Therefore, m∠ BDE = 36° since the sum of the measures of ∠ EDB and ∠ BDA must be 180°.

Δ EDB is also isosceles, so m ∠ DBE = m ∠ BED = 72°.

Therefore, Δ EDB is Golden and DE = Ø.

We can use this information to develop another way to cut a line segment, this time the radius of a circle, into Ø proportions.

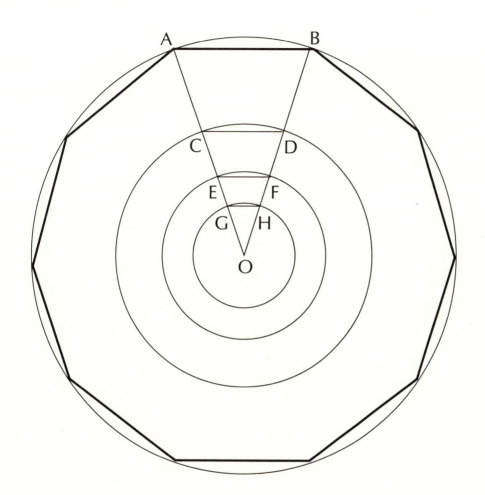

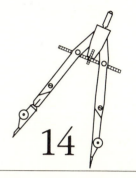

14

Divide the Radius of a Circle into Ø Proportions

Given: a regular decagon with side $\overline{AB}$ inscribed in circle O

1. Draw a circle with center O and radius equal to the length of $\overline{AB}$.

2. Label points of intersection with $\overline{OA}$ and $\overline{OB}$ C and D, respectively, and draw $\overline{CD}$.

Note that triangle OAE contains the framework for the Lute of Pythagoras (see Book 1, Universal Patterns).

3. Repeat with Δ OCD to get E and F, and with Δ OEF to get G and H.

Now both radii $\overline{OA}$ and $\overline{OB}$ are divided into Ø-proportions.

Again, why does this work?

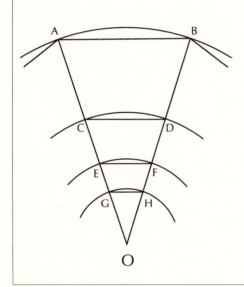

Fig. 3.30 The radius of a regular decagon is divided into Ø proportions by the length of the side.

Proof:

When concentric circles are drawn, the figure encompassed between the two radii is a series of nested similar Golden Triangles.

We already know that $\dfrac{OA}{AB} = \dfrac{OC}{CD} = \dfrac{OE}{EF} = \dfrac{OG}{GH} = Ø$.

We also know that AB = OC, CD = OE and EF = OG.

By substituting into the first equation, we get:

$\dfrac{OA}{OC} = Ø$, so C is the Golden Section of $\overline{OA}$.

$\dfrac{OC}{OE} = Ø$, so E is the Golden Section of $\overline{OC}$.

$\dfrac{OE}{OG} = Ø$, so G is the Golden Section of $\overline{OE}$.

Therefore, $\overline{OA}$ is divided into Ø-proportions. Similarly, it is true for $\overline{OB}$.

Below:
Circular armature in Roman paving tiles.

We have already mentioned that the circumference of a circle relates directly to the irrational number π (roughly equivalent to 3.142) by the formula C = πd, where d is the diameter. Scholars have noticed that there also exists a relationship between π and our other irrational friend, Ø. Although it cannot be described by an equation, as above, it is a "ballpark" approximation.

$$\left(\frac{\pi}{4}\right)^2 \approx \left(\frac{3.142}{4}\right)^2 \approx \frac{9.872}{16} \approx .617 \approx \frac{1}{Ø}.$$

In actuality, $\dfrac{1}{Ø}$ (the reciprocal of Ø) is closer to .618.

Another set of polygons may be obtained from the inscribed ones just discussed. For example, begin with the points that would determine an inscribed pentagon. Find the midpoints of the sides, as before, and draw radii through those points. With the addition of a smaller concentric circle, we have the framework for what could be called a "star polygon." The fatness of the star depends upon the radius of the inner circle as illustrated by the three different ones in Fig. 3.31. This concept can be applied to any regular polygon.

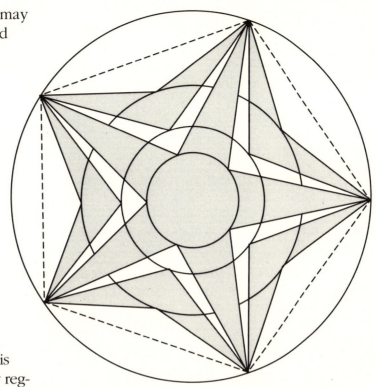

Above:
Fig. 3.31 Star decagons developed from a regular pentagon.

Below;
Fig. 3.32 The length of the Golden Rectangle is equal to the side of the inscribed regular hexagon.

Circles in Relation to Dynamic and Golden Rectangles

Since the Golden Ratio is the essential thread that weaves together the concepts under consideration, let us examine the relationship of Ø to circles. At first glance there seems to be no apparent connection between them. Yet, if we choose to draw a circle whose radius is equal to the length of a given Golden Rectangle, three little surprises pop up. The first two are directly related to the previously discussed ideas.

1. The length of the rectangle is the length of a side of the inscribed hexagon (Fig. 3.32). This is another way of saying the radius of the circle determines a side of the inscribed hexagon.

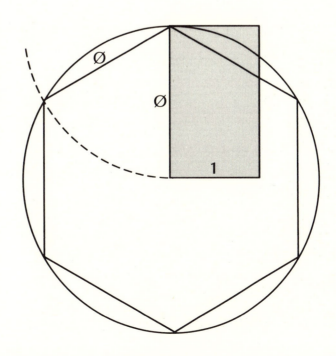

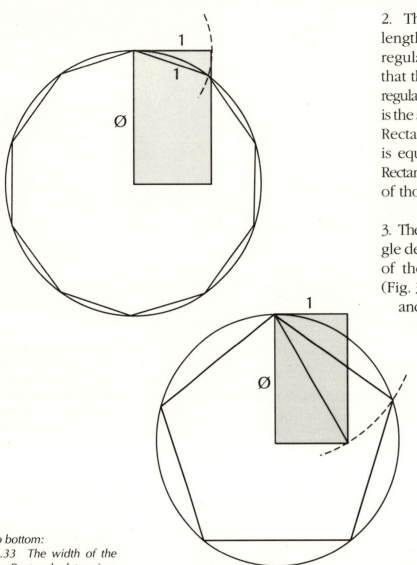

2. The width of the rectangle is the length of a side of the inscribed regular decagon (Fig. 3.33). Recall that the ratio of the radius to a side of a regular decagon is Ø. Here, the radius is the same as the length of the Golden Rectangle and a side of the decagon is equal to the width of the Golden Rectangle. Thus, the ratio of the lengths of those sides is Ø.

3. The diagonal of the Golden Rectangle determines the length of the side of the inscribed regular pentagon (Fig. 3.34). While this is visually clear, and perhaps affords an easier construction of a regular pentagon, we should again ask why it works. If we accept the construction of the regular pentagon inscribed within a circle as given in Construction 11, Book 1, *Universal Patterns*, we can quickly see the relationship between the Golden Rectangle and the pentagon (Fig. 3.35).

Top to bottom:
Fig. 3.33 The width of the Golden Rectangle determines the length of the side of the inscribed regular decagon.

Fig. 3.34 The diagonal of the Golden Rectangle determines the length of the side of the inscribed regular pentagon.

Fig. 3.35 Since A is the midpoint of $\overline{OD}$, and C is determined by an arc with radius AB, then CDEF is a Golden Rectangle. Since ODEB is a square, it follows that COBF is also a Golden Rectangle, one whose length is the radius of the circle. Recall, then, that it is BC that determines the length of the arcs to be cut successively around the circle to construct the regular pentagon, and $\overline{BC}$ is the diagonal of the Golden Rectangle.

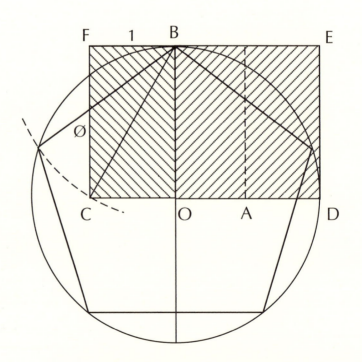

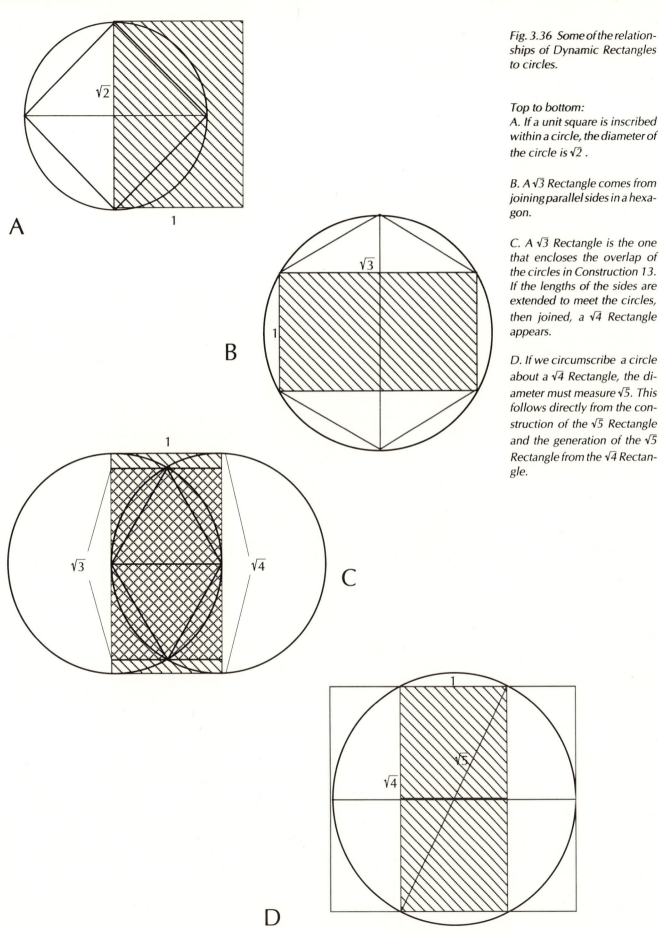

Fig. 3.36 Some of the relationships of Dynamic Rectangles to circles.

Top to bottom:
A. If a unit square is inscribed within a circle, the diameter of the circle is $\sqrt{2}$.

B. A $\sqrt{3}$ Rectangle comes from joining parallel sides in a hexagon.

C. A $\sqrt{3}$ Rectangle is the one that encloses the overlap of the circles in Construction 13. If the lengths of the sides are extended to meet the circles, then joined, a $\sqrt{4}$ Rectangle appears.

D. If we circumscribe a circle about a $\sqrt{4}$ Rectangle, the diameter must measure $\sqrt{5}$. This follows directly from the construction of the $\sqrt{5}$ Rectangle and the generation of the $\sqrt{5}$ Rectangle from the $\sqrt{4}$ Rectangle.

The complexity of Gothic architecture belies the underlying simplicity of the structures where the
circle and inscribed polygons control
the design. One of the fundamental
diagrams transmitted from Gothic
Master to Master provided a ground
plan and a key for three-dimension-
al construction. It consisted of the
pentagon and decagon placed within the circle of orientation. Sometimes the circle was divided into as
many as twenty parts, but in such
cases the harmony of the pentagon
and Golden Ratio was preserved.
Fig 3.37 shows an example of such
an underlying structure.

In other cultures the inscribed

square is of paramount importance.
When the circle is joined with the
square, their juxtaposition suggests
the relationship of Heaven to Earth.
While the circle represents everywhere,
the square points to the four primary
directions of North, South, East and
West. It stands for the four seasons,
the four elements of earth, air, fire
and water, and the four stages of
human development: childhood, youth,
maturity, old age. It suggests the limited as opposed to the circle's function as infinity. Visually, this relationship is quintessentially expressed
through the Tibetan mandala and the
American Southwest Indians' sand-
painting—vehicles for meditation.
Jungian psycholo-
gy of the Twenti-
eth Century also
uses it as a mean-
ingful form, for it is
a means of focus-
ing upon integra-
tion, both per-
sonal and
cosmic.

Below:
*Fig. 3.37 This diagram con-
sists of two overlapping pentagons together with their pentagrams which provide key
points for constructing Gold-
en Rectangle ABCD and the
smaller pentagrams within the
inner circle.*

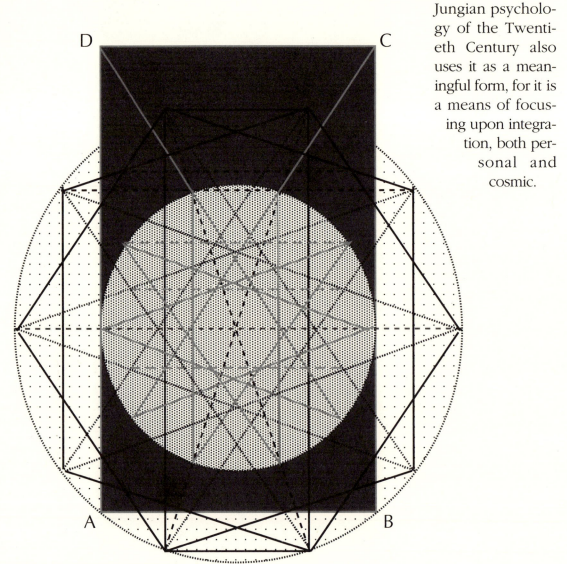

Gina Halpern. The Crown Chakra or The Many Petaled Lotus. *Mixed media.*

This image was created as part of a series of eight meditation paintings designed to be used in the healing process. Each painting centers on one color of the spectrum, and utilizes the complementary color to develop an harmonious relationship. These images correspond to the notes of the musical scale which were formed into the eight movement musical composition, A Rainbow Path, by composer Kay Gardner. The entire project was based on the theory of Pythagoras who believed that the notes of the musical scale corresponded to the colors of visible light spectrum, and its connection to the Eastern chakra system. The Image of the Crown Chakra, or Many Petaled Lotus, relates to the concept of the opening of the mind to the presence of Divine or Universal order. In contemplating the power of this concept, the vision of the Golden Rectangle and its beautifully unfolding spiral emerged as the perfect symbol. By using the two rectangles in a mirror image which both projected inward and blossomed outward, I attempted to draw the viewer into a state of peace.

Gina Halpern

Problems

[1] Use the concept of straight line or circular armatures to design a circular pattern not found in the text.

[2] What would the radius of each circle need to be if you wanted to pack 25 circles within an eight inch square in the square arrangement described on page 108?

[3] Divide a circle into 21 congruent arcs.

[4] Draw a circle with radius of at least 3 inches and construct an internal octad (eight circles.) Do the same for an internal nonad (nine circles.)

[5] Using geometric construction, analyze the tracery pattern shown in Fig. 3.38. Hint: start with an internal triad and develop a structure using six overlapping circles the size of the original three.

[6] Construct a family of regular polygons having an edge length of one inch.

[7] Inscribe a regular octagon in a circle and create three different star polygons using the octagon as parent.

[8] Photocopy an example of a Rose window and analyze its structure. Look for connections to Fibonacci numbers, Golden Rectangles or dynamic figures.

Projects

1 On acetate, photocopy one of the grids from Chapter 1. Do the same with one of the armatures of the circle from this chapter. Superimpose the two and make a tracing of the composite image. Eliminate whatever lines or areas that confuse your image. Do one of the following.

 a. Use only black marks on white paper to create your structure.

 b. Use black and white areas on gray paper to create your structure.

 c. Use colors on black paper to create your structure.

2 Using a tondo painting from the history of art as your source, create a version of it in collage.

3 Using the concepts of armature, and close packing in square or isometric arrangements, create a pattern which will be used as the basis for designing a family crest. Execute in materials of your choice.

4 Use the subdivisions found in Problem 1 to develop a circular composition. Use it to transform an old umbrella.

5 Design four different dinner plates using circular armatures. Design them as a set using the concept of lunar cycles.

6 Embellish a circular fan form with a natural design developed from one of the armatures described in this chapter.

7 Sew a circular skirt, cape, poncho or dress with a decorative pattern that brings together circular sun and moon images.

8 Create a two or three dimensional composition using different star polygons that have the same parent.

9 Using your knowledge of the relationship between Dynamic Rectangles and circles, create a composition whose underlying structure consists of the union of a circle, or circles, and one of the Dynamic Rectangles.

10 Use one of the results of Problem 4. Enlarge to a diameter of at least 18". Use illustration board as "leading" and set colored acetate behind to create a "stained glass" window.

11 Using the pattern from Problem 5, create a fantasy flower or leaf that has the same basic armature.

12 Design and execute a circular contemporary "shield" that acts as symbolic rather than physical protection.

13 Use the concept of the Yin and Yang symbol to create a piece in two or three-dimensions that retains the original flavor of the symbol but transforms it in some way.

14 Design and execute a healing mandala based on the concept of *squaring the circle*. Research the roots of this geometric problem.

15 Research the ground plans of medieval cathedrals. Through visual documentation, describe their underlying geometry. Discuss in writing why you think the creators chose the particulars of these buildings. What is the relationship between the forms and the prevailing world view at that time.

Below:
Student work. Linda Maddox.
Technical pen on paper.

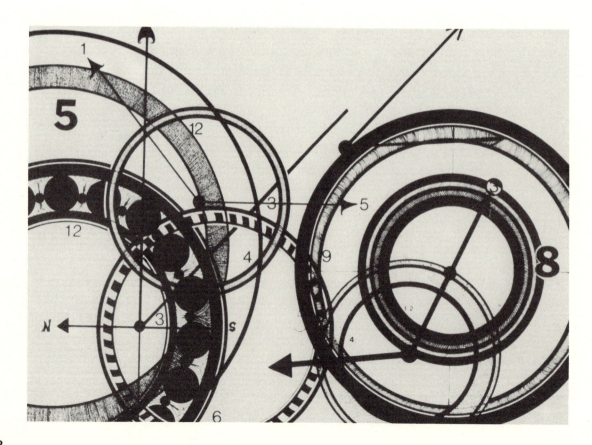

Further Readings

Arguelles, Jose and Miriam. *Mandala*. Berkeley: Shambala Publications, 1972.

Billings, Robert. *Power of Form Applied to Geometric Tracery*. Edinburgh and London: W. Blackwood and Sons, 1851.

Cowen, Painton. *Rose Windows*. London: Thames and Hudson, Ltd., 1990.

Critchlow, Keith. *Islamic Pattern*. New York: Schocken Books, 1976.

Foy, Sally. *The Grand Design: Form and Color in Animals*. Englewood Cliffs. N.J: Prentice Hall, 1982.

Lawlor, Robert. *Sacred Geometry*. New York: The Crossroad Publishing Co., 1982.

Leapfrogs. *Curves*. Stradbroke, Diss, Norfolk: Tarquin Publications, 1982.

Pierce, Peter. *Structure in Nature as a Strategy for Design*. Cambridge: The MIT Press, 1978.

Pedoe, D. *Circles: A Mathematical View*. New York: Dover Publications, 1979.

Shaw, Sheilagh. *Kaleidometrics*. Stradbroke, Diss, Norfolk: Tarquin Publications, 1981.

Wilhelm, Richard (translation and explanation by) *The Secret of the Golden Flower*. New York: Harcourt, Brace & World, Inc., 1962.

4 Tiling the Plane

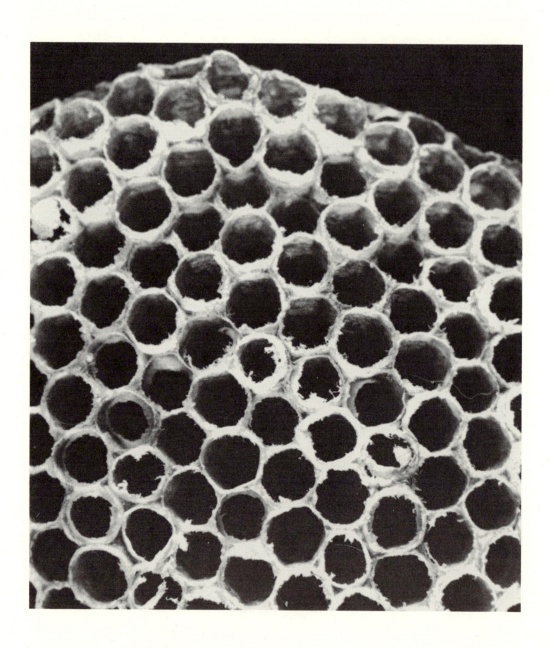

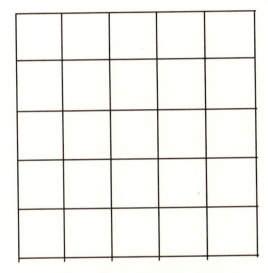

Pattern, like number, is one of the fundamental conditions of existence and is likewise a vehicle of archetypes.
Keith Critchlow
Islamic Patterns

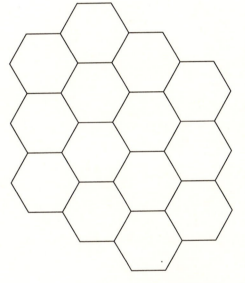

Left:
Fig. 4.1 The three possible ways to cover the plane with regular units.

Below:
Fig. 4.2 According to Keith Critchlow in his book, Islamic Pattern, these three essential tile units also act as symbols. The square refers to the earth and materiality, the triangle to human consciousness and the hexagon to Heaven. The diagram symbolizes the relationship of the elements to each other and to the universe which is represented by the circle.

Tilings have an intimate relationship with circles, grids and symmetry, and, in fact, sometimes it is difficult to separate the concepts. In the previous chapter we saw that circles can be tightly packed together, but always in such a way that there are small leftover portions of the plane. If, however, the circles are pushed hard against each other, the more they are compressed the more they will be forced into an hexagonal structure which permits no leftover areas. When a flat surface must be completely covered with a number of single *regular* units, there are only three options available: the aforementioned hexagon, the square, and the equilateral triangle (Fig. 4.1).

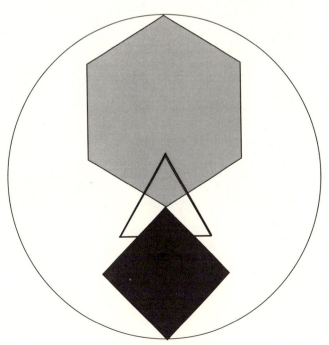

> The world is so full of a number of things
> I'm sure we should all be as happy as kings.
>
> Robert Louis Stevenson

it is only within the last hundred years that the underlying mathematics has been a concern for serious investigation, with much of the research taking place in the latter half of the twentieth century. It is not our intent here to explore the mathematical intricacies of the subject. Grunbaum and Shephard have written a detailed study in their book, *Tilings and Patterns,* and we refer you to this excellent source. Our desire is to provide just enough background so that you can play forever.

Again, let our inspiration come from Nature. Natural structures are always economical, using the least amounts of energy and materials. Consider the honeycomb. A cross-section reveals an hexagonal tiling. Why not a square or triangular one? In actuality, the structure is a system of compressed wax tubes. Cylinders, such as these, when stacked, will arrange themselves isometrically. When made of malleable material and pressed together, the hexagonal shape emerges. This configuration provides structural stability, and allows for maximum enclosure with minimal material.

Tiling is the process of covering the plane with closed units which are joined so that there are no gaps or overlaps. A tiling is the resulting pattern, so that the word is both a verb and a noun. The Latin word for tile is *tesserae*, and thus sometimes a tiling is called a tessellation.

At first appearance the concept of tiling seems deceptively simple. On further investigation, however, the subject becomes fascinatingly complex. In spite of the fact that tiles have been used for many centuries,

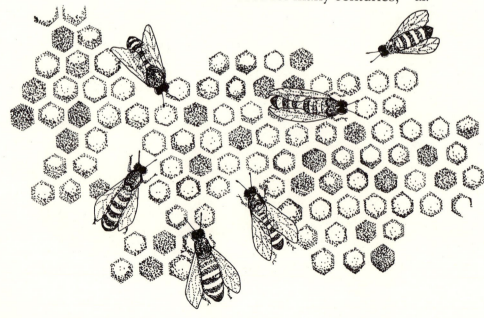

Looking again at the planar situation, the hexagonal array is the simplest regular tiling since each vertex is shared by three rather than four or six sides (Fig. 4.3). This minimizes weakness. The three-way joint is a common natural planar division irrespective of the size of the surface. Three-way joints of 120° result most often in situations of compression, such as in the aggregation of cells and bubbles. This same type of structure is evident in the patterns on the skins of many animals. Natural cracking also forms tilings in which polygonal shapes occur in a more or less systematic pattern. When stress becomes too great on an elastic surface, it cracks. New cracks form at right angles to old ones so that, in this case, the three-way joint contains 90° and straight angles as well as the 120° ones. This happens when mud dries, paint peels, and, in some instances, when pottery glazes are fired.

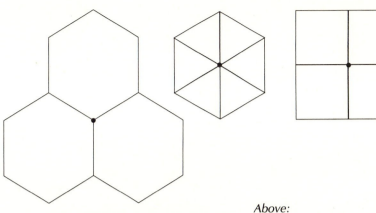

Above:
Fig. 4.3   Three tiles share a vertex in the hexagonal tiling, four in the square one and six in the triangular one.

Below:
Tilings are suggested by natural forms:
    Giraffe skin
    Pottery glaze
    Cracking paint
    Pangolin scales

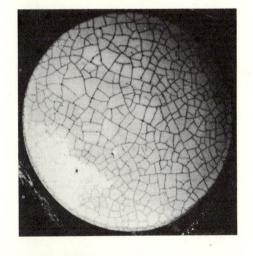

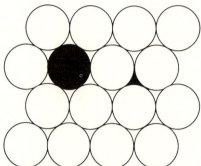

By our previous definition, the two fundamental arrays of closely packed circles do form tilings. In Fig. 4.4 the two types of tiles formed by each array are shaded. If the centers of the circles are connected with line segments, square and isometric grids emerge. An hexagonal grid is formed when key centers are not used (Fig. 4.5). These grids, in turn, are tilings of the most basic kind wherein a single regular polygonal unit is used. They are known as the **regular tilings**.

Each of the seventeen plane groups of symmetry discussed in Chapter 2 involves manipulation of a unit that, when translated, results in a tiling. Not all tilings are symmetric but all the plane groups of symmetry are tilings in which the unit is the underlying tile. For any given plane group, the tiling may be either content or substructure, depending upon the choice of the motif and the manner in which it relates to the unit.

Above:
Fig. 4.4   Circular tilings in square and isometric formats. The two types of tiles in each case are shaded.

Below and right:
Fig. 4.5 Connecting the centers of the circles results in square, triangular and hexagonal grids. Shading delineates the tile units.

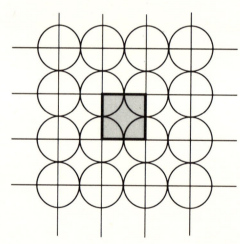

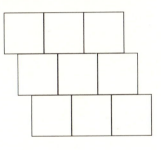

Above:
Fig. 4.6   One of the infinite number of ways to tile with only a square.

How to Begin

When confronted with a large quantity of complex information, setting limits to both the questions and conditions considered helps an individual's understanding. This leads to manageability and control which is an essential part of the mathematical process. Mathematically, a definition sets limits in order to eliminate ambiguities, while offering clarity and promoting consistency. Sloppiness at these seemingly small beginnings leads to extremely poor mental housekeeping later on. Our discussion here will barely blow away some of the cobwebs from the surface in hopes that you will gain enough insight to be able to use it for your own purposes.

Given our initial definition, it is possible for tiles to assume many different shapes. Therefore, one of the first limits we will set is to work with only polygonal ones. This is a huge limitation since there are infinitely many other shapes to consider, and, in fact, you will see ways to extrapolate some of these later on in the chapter.

Let us further restrict our tiles to those formed by regular polygons. These are good ones to start with because they allow us to answer some basic questions about how a portion of the plane can be tiled.

Suppose we begin with a square. In addition to the regular tiling we have already seen, there are infinitely many other ways to tile with only congruent square tiles. One of these is shown in Fig. 4.6.

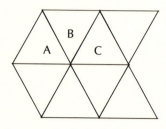

Above:

Fig. 4.7 In an edge-to-edge tiling, tiles may share a vertex (A and C) or an edge and two vertices (B and C).

Given that there are infinitely many points along the side of a square, a row of squares can be shifted any number of ways. With so many possibilities there is still too much to consider. We need a further restriction; one, which when using squares, will take us back to the aforementioned regular tiling. The restriction is that if two tiles intersect, their intersection is either (1) a common vertex, or (2) a common side with two common vertices (Fig. 4.7). Tilings that have this property are called **edge-to-edge** tilings. If we re-examine Fig. 4.6 we see that tiles share a part of a side and have no common vertices, and, therefore, this tiling is excluded by our restriction.

We already know that there are only three regular tilings. Therefore, our next question is, "Which regular polygonal tiles can be used in combination to tile a portion of the plane?" Since, by our restriction, vertices must be shared, let us start there. We know that 360° surround each point in the plane. So the sum of the angles in the polygons surrounding a vertex must be 360°. Fig. 4.8 shows that this is certainly the case for the regular tilings. Fig. 4.9 gives the measure of the interior angles for some of the regular polygons. The chart could be continued by using the formula

$$m = \frac{180(n-2)}{n}$$ where m is the measure of an interior angle and n is the number of sides in the polygon.

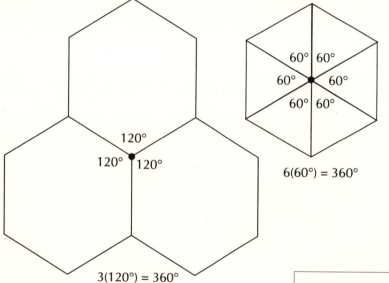

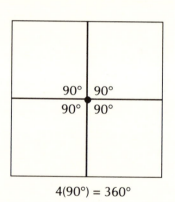

Above:

Fig. 4.8 The sum of the angles surrounding a vertex must total 360°.

Right:

Fig. 4.9 Measure of interior angles of regular polygons.

Number of sides	Measure of interior angles
3	60°
4	90°
5	108°
6	120°
7	128 4/7°
8	135°
9	140°
10	144°
11	147 3/11°
12	150°

The next task is to see which combinations give an angle sum of 360°. Careful examination shows that if we limit ourselves to the polygons in Fig. 4.9, the possibilities are restricted to the twelve in Fig. 4.10. There are actually twenty-one possibilities, but the remaining ones include the use of polygons having 10, 15, 18, 20, 24 or 42 sides, most of which are not especially practical for design purposes. Using the formula and the chart, can you discover the others?

For each of the twelve cases, Fig. 4.10 shows what we will call a **vertex net**. A numerical shorthand, called the **order** of the net, is also given. The number indicates the number of sides in the polygon, and the order in which the numbers are arranged represents either a clockwise or counterclockwise trip around the vertex. The numerical names are not unique (See Fig 4.11).

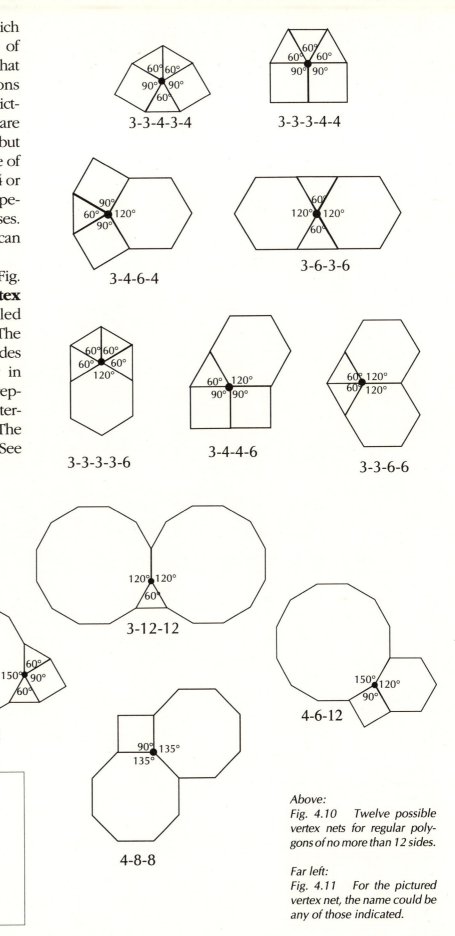

3-3-4-3-4

3-3-3-4-4

3-4-6-4

3-6-3-6

3-3-3-3-6

3-4-4-6

3-3-6-6

3-3-4-12

3-4-3-12

3-12-12

4-6-12

4-8-8

4-6-4-3
6-4-3-4
4-3-4-6
3-4-6-4

Above:
Fig. 4.10   Twelve possible vertex nets for regular polygons of no more than 12 sides.

Far left:
Fig. 4.11   For the pictured vertex net, the name could be any of those indicated.

Our next question is, "How can we use multiples of the vertex nets to tile the plane?" If our only concern is for tiling, then there are limitless ways to use them. However, if we also want to have uniformity, continuity and pattern (as would a designer of fabric, floor covering, wallpaper, etc.) we then have fewer choices.

Let us begin with tilings which use multiples of a single vertex net. This means that every vertex of the tiling is surrounded by polygonal tiles that are alike in kind, order (disregarding direction) and number. In Fig. 4.12, three tilings using triangles and squares are shown. Does each fit the aforementioned restrictions? If we choose several vertices at random, we get a better idea of how to answer this question. Check the vertices marked in diagram A. Looking only at the tiles surrounding a vertex, we see that in each case the net is of the order 3-3-4-3-4. This is also the case in B where the net is of the order 3-3-3-4-4. However, one vertex in C is of order 3-3-3-4-4 while another is of order 3-3-4-3-4. Thus, this tiling does not satisfy our criterion.

Below:
Fig. 4.12 Three tilings using squares and triangles. A and B are uniform, C is not.

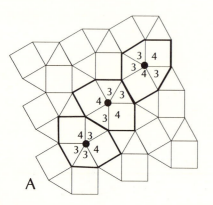

A

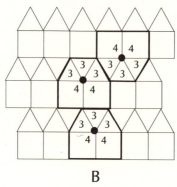

B

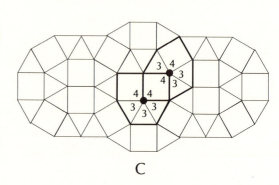

C

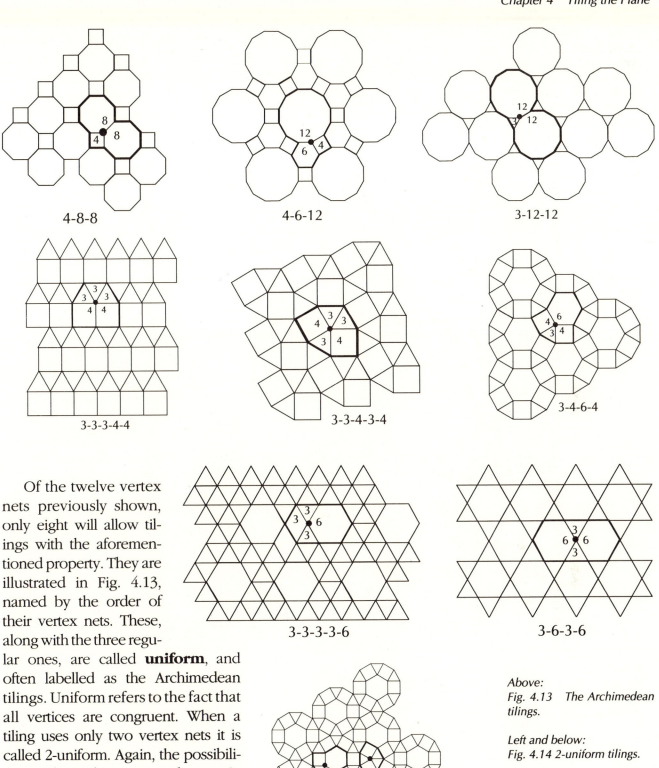

4-8-8

4-6-12

3-12-12

3-3-3-4-4

3-3-4-3-4

3-4-6-4

3-3-3-3-6

3-6-3-6

Of the twelve vertex nets previously shown, only eight will allow tilings with the aforementioned property. They are illustrated in Fig. 4.13, named by the order of their vertex nets. These, along with the three regular ones, are called **uniform**, and often labelled as the Archimedean tilings. Uniform refers to the fact that all vertices are congruent. When a tiling uses only two vertex nets it is called 2-uniform. Again, the possibilities are limited in structure but not in design potential. Fig. 4.14 gives examples of 2-uniform tilings. Notice they are named by both vertex net orders. Other tilings may use 3, 4, 5, . . ., k vertex nets, in which case they are called 3-uniform, 4-uniform, 5-uniform, . . ., k-uniform, where *k* is any natural number.

Above:
Fig. 4.13   The Archimedean tilings.

Left and below:
Fig. 4.14 2-uniform tilings.

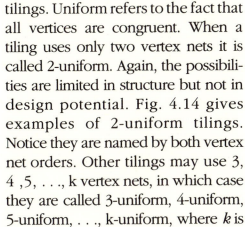

3-4-6-4 3-3-4-3-4

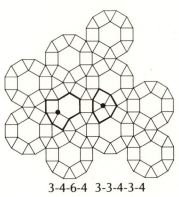

3-3-3-3-3-3 3-3-3-4-4

With this bare bones introduction to the subject of tiling, we now wish to examine some which deviate from these very tight limits. Rather than repeating information found in the many good books on tilings, we will explore those which grow out of our interest in natural structure and harmonic relationships related to universal patterns.

Above and right:
Fig. 4.15 Tilings with Triangles of Price.

Below:
Fig. 4.16 Tilings with Golden Triangles. Three examples abide by the edge-to-edge rule; two do not.

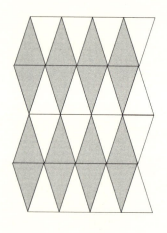

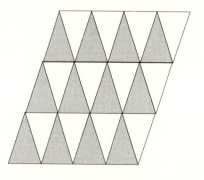

Triangular Tilings

It is known that all triangles tile. This gives the designer countless choices beyond the uniform tilings that use equilateral triangles. Here we would like to look at some of the possibilities afforded by the Special Triangles (Chapter 3, Book 1, *Universal Patterns*). Fig. 4.15 uses the Triangle of Price, but all the Special Right Triangles can be used to create tilings such as these.

The Golden Triangle, too, can be used to create various tilings. If the edge-to-edge rule is suspended, rows of triangles may be translated. For example, the vertex angles of one row may fall on the Golden Cut of the bases of those in the previous row. Or triangles may be arranged so that the rows appear twisted (Fig. 4.16).

So far we have looked at triangular tilings in which all tiles are congruent to a single one called the **prototile**. Remember that congruence requires the same size and shape, but the symmetry operation of reflection is allowed. Tilings which have a single prototile are called **monohedral**.

A tiling can be created using similar prototiles of different sizes. In Fig. 4.17, A shows a **dihedral** tiling (two prototiles) whereas B shows a **trihedral** one (three prototiles). A tiling which has *n* prototiles is called **n-hedral**.

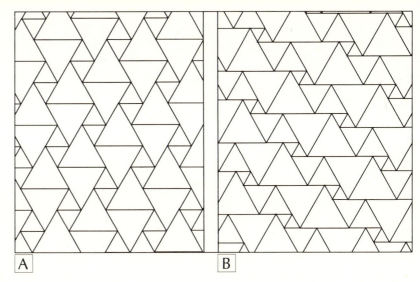

Above:
Fig. 4.17   Dihedral and trihedral tilings using similar prototiles.

Below:
Fig. 4.18   A uniform trihedral tiling in which two prototiles are equilateral triangles and one is an isosceles triangle.

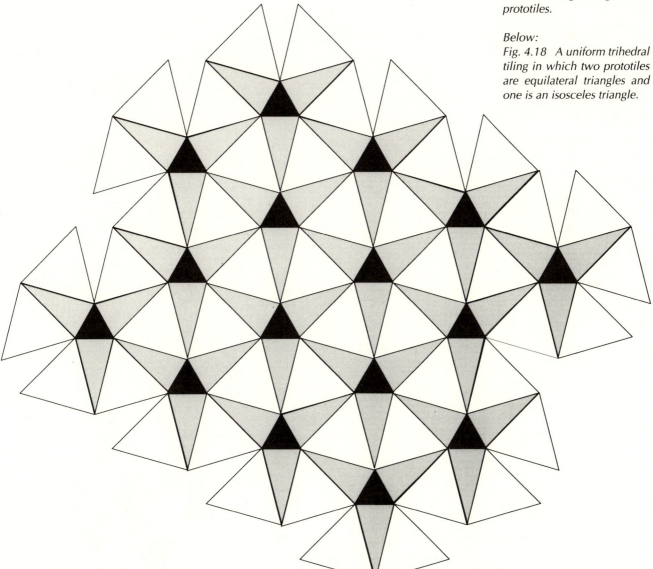

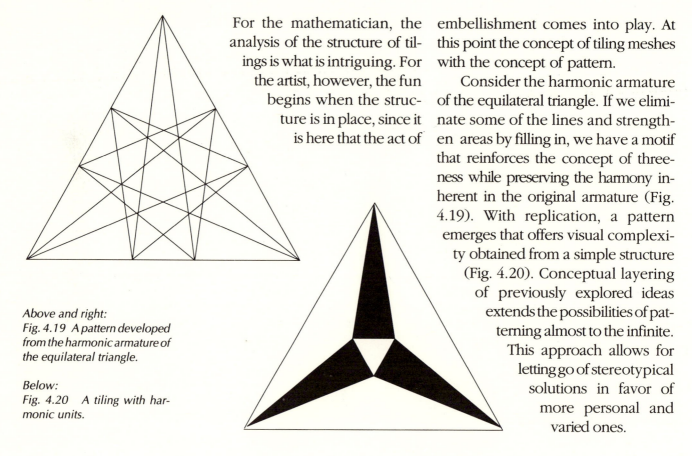

For the mathematician, the analysis of the structure of tilings is what is intriguing. For the artist, however, the fun begins when the structure is in place, since it is here that the act of embellishment comes into play. At this point the concept of tiling meshes with the concept of pattern.

Consider the harmonic armature of the equilateral triangle. If we eliminate some of the lines and strengthen areas by filling in, we have a motif that reinforces the concept of threeness while preserving the harmony inherent in the original armature (Fig. 4.19). With replication, a pattern emerges that offers visual complexity obtained from a simple structure (Fig. 4.20). Conceptual layering of previously explored ideas extends the possibilities of patterning almost to the infinite. This approach allows for letting go of stereotypical solutions in favor of more personal and varied ones.

Above and right:
Fig. 4.19 A pattern developed from the harmonic armature of the equilateral triangle.

Below:
Fig. 4.20 A tiling with harmonic units.

Hexagonal Tilings

Since the hexagonal grid comes from the triangular one, tilings with triangles and hexagons become inextricably bound. Unlike triangles, not all hexagons tile. However, the limits of having three pairs of parallel and congruent opposite sides guarantee that an hexagon will tile (Fig. 4.21).

If an hexagonal prototile possesses bilateral symmetry, where the axis of reflection is through a pair of congruent opposite sides, it will also tile (Fig. 4.22). We refer the reader, again, to the work of Grunbaum and Shephard for more detailed investigation.

Fig. 4.23 shows how a dihedral tiling can be generated using hexagonal prototiles. It could be developed directly from the isometric grid by **truncating** the triangular units while preserving the bilateral symmetry in each.

Equilateral triangles and regular hexagons combine in Fig. 4.24 to form a dihedral tiling where the pattern structure changes depending on the ratio of the lengths of the sides of the two prototiles. The overall appearance of the surface can be completely transformed with the addition of surface pattern.

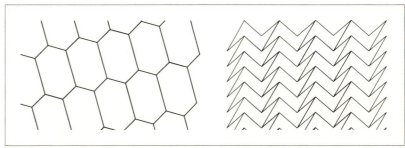
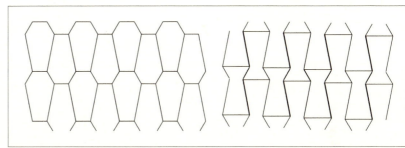
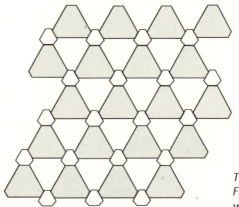
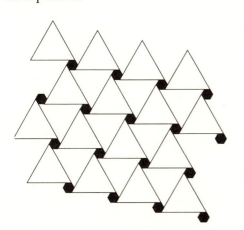

Top to bottom:
Fig. 4.21 Hexagonal tilings in which the prototile has three pairs of parallel and congruent opposite sides.

Fig. 4.22 Hexagonal tilings in which the prototile has bilateral symmetry with the axis through parallel sides.

Fig. 4.23 Dihedral hexagonal tiling.

Fig. 4.24 Dihedral tilings with regular hexagons and equilateral triangles.

Right:
Fig. 4.25 Tilings using irregular quadrilaterals.

Center right:
Fig. 4.26 Lutes of Pythagoras in a tiling.

Center left:
Fig. 4.27 Dihedral tilings with pairs of parallelograms.

Lower left:
Mark McGrade. Student work. Acrylic paint on toned paper.

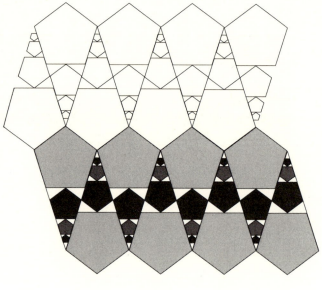

Quadrilateral Tilings

Any quadrilateral may be a prototile for a monohedral tiling. We urge you, therefore, to explore the potential for tiling with Dynamic and Ø-Family Rectangles and Parallelograms, Lutes of Pythagoras, as well as irregular quadrilaterals. Take pattern permutations within the tiles into consideration for a personal touch.

Another useful property of tiling with quadrilaterals is that dihedral tilings may be obtained from any two rectangles, or any pair of parallelograms, whose angles are congruent (Fig. 4.27). In a practical sense, this allows for great variety using readily available tile shapes such as those found in ceramics or wood parquet. Again, we urge you to explore combinations that preserve harmony.

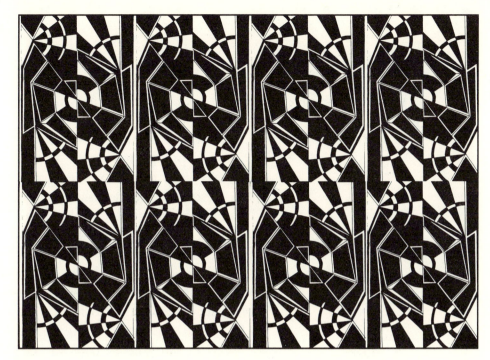

Left:
Tiling with student created Golden Rectangle unit. Damon Dyer. Ink on paper.

Below:
Fig. 4.28 A nonuniform tiling with rotational symmetry.

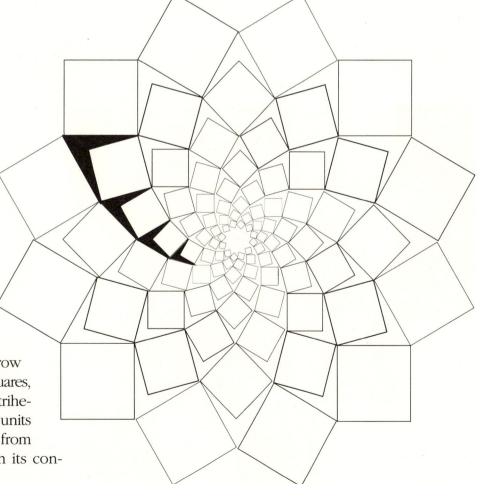

Each of the aforementioned tilings with quadrilaterals has translational symmetry. Now let us examine one that has rotational symmetry, and scaling, instead (Fig. 4.28). With the exception of the center, it consists of two types of tiles, squares and the arrow shaped ones between the squares, but it is neither dihedral nor trihedral since the size of the units increases as we move out from the center. Information on its construction follows.

145

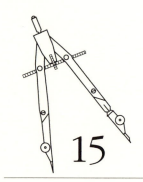

15

Construct a
Tiling
Consisting of
Spiraling
Squares

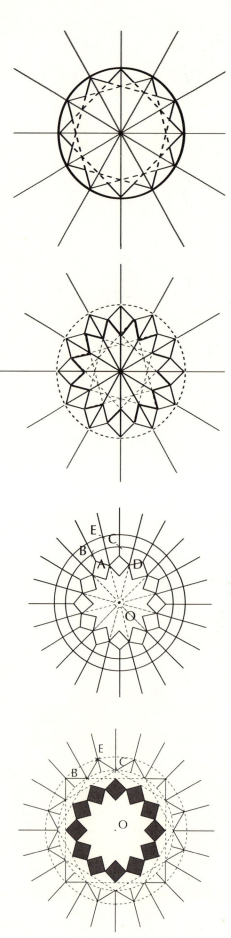

1. Divide the circle into 12 congruent arcs (Central angles measure 360°/12 = 30°). Extend the radii through the endpoints of the arcs. Draw three overlapping squares by connecting every *third* point on the circle. Only the outer portions will be part of the finished tiling, so the parts indicated with dotted lines may be erased.

2. Connect every *fourth* one of the innermost vertices of the star 24-gon created in Step 1 to form a ring of twelve small squares. Again, the dotted lines indicate the parts not needed.

3. Draw and extend the diagonals through the common vertices of the twelve small squares. These, together with the original diagonals, provide the reference points for vertices of the subsequent rings of squares. Open the compass to measure the diagonal of one of the squares $\overline{AD}$. With the metal tip on A, cut arcs that intersect the diagonals on either side of A. Label intersection points B and C. Without changing the setting, place the metal tip on B and cut an arc that intersects the diagonal through A. Label the intersection point E. Draw a circle with radius OE and another with radius OB.

4. Again, draw the star 24-gon by using a series of three overlapping squares, as in Step 1. Notice that the inner circle (with radius OB) and the radiating diagonals now determine the points of intersection of the sides of the three squares. This point the dotted portions of the sides are no longer necessary to draw. They will not be shown for subsequent rings of squares.

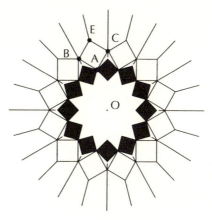

5. Form the remaining sides of the squares by joining common vertices on the outer ring to common vertices on the inner ring. ($\overline{AB}$ and $\overline{AC}$, for instance).

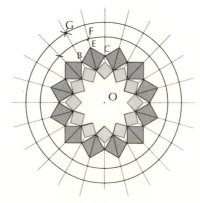

6. Using diagonal $\overline{BC}$ and the technique described in Step 3, draw circles with radii OF and OG.

7. Repeat Steps 4 and 5 to complete the third ring of squares.

8. Continue to add rings of squares in this manner until the tiling is of the desired size.

Now the tiling is complete.

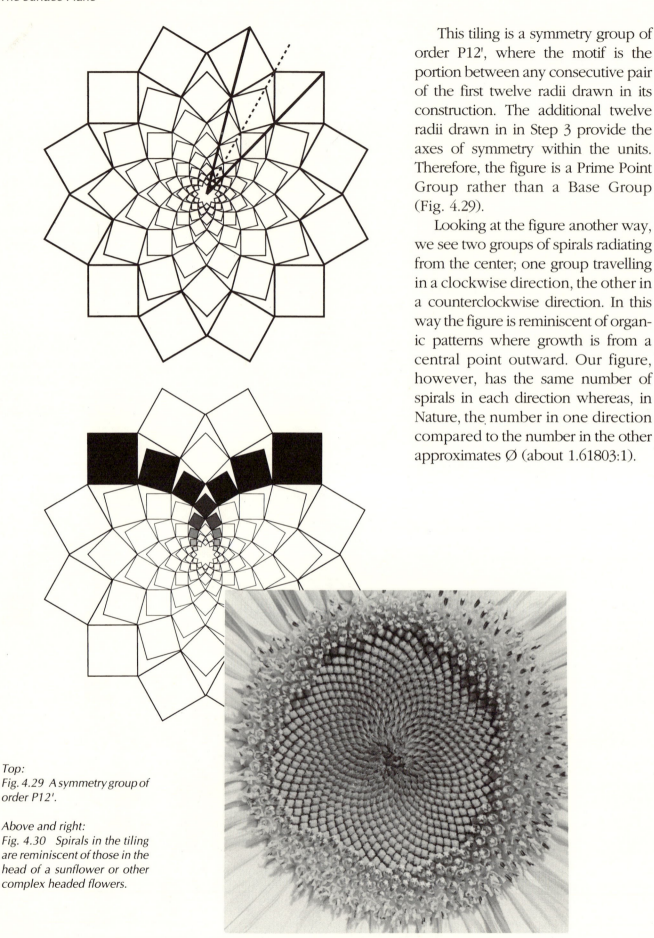

This tiling is a symmetry group of order P12', where the motif is the portion between any consecutive pair of the first twelve radii drawn in its construction. The additional twelve radii drawn in in Step 3 provide the axes of symmetry within the units. Therefore, the figure is a Prime Point Group rather than a Base Group (Fig. 4.29).

Looking at the figure another way, we see two groups of spirals radiating from the center; one group travelling in a clockwise direction, the other in a counterclockwise direction. In this way the figure is reminiscent of organic patterns where growth is from a central point outward. Our figure, however, has the same number of spirals in each direction whereas, in Nature, the number in one direction compared to the number in the other approximates Ø (about 1.61803:1).

Top:
Fig. 4.29 A symmetry group of order P12'.

Above and right:
Fig. 4.30 Spirals in the tiling are reminiscent of those in the head of a sunflower or other complex headed flowers.

Recall that, in the construction, the diagonal of a square in one ring was used to determine the side of a square in the subsequent ring. Therefore, there is harmonic growth among the squares of a particular spiral. The growth factor for the sides of the squares is $\sqrt{2}$ while the areas of any consecutive pair of squares is 2:1 (Fig. 4.31).

In the abstract, this tiling is capable of infinite expansion. In Nature, growth is limited by constraints of space and particular needs of an organism. Growth of this pattern is limited by the size of the surface to be tiled and the time and patience of the designer.

Since this figure is based on a pattern of three overlapping squares, we can cut the tiling along the sides of any one of the squares to get a prototile for other tilings. With the addition of color or value changes and the symmetry operations of rotation and reflection, patterns of amazing complexity can be gotten from the simplest of regular tilings— the square grid.

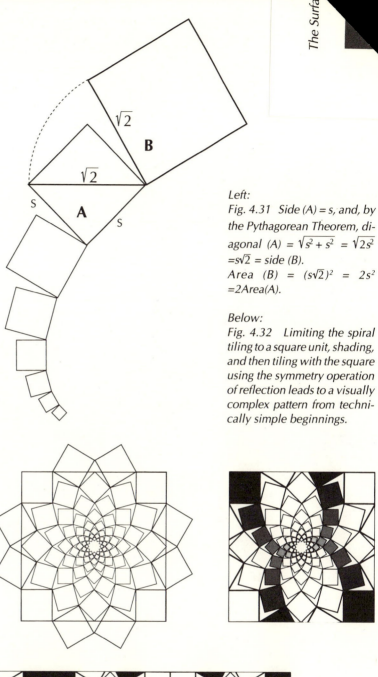

Left:
Fig. 4.31 Side (A) = s, and, by the Pythagorean Theorem, diagonal (A) = $\sqrt{s^2 + s^2}$ = $\sqrt{2s^2}$ =$s\sqrt{2}$ = side (B).
Area (B) = $(s\sqrt{2})^2$ = $2s^2$ =2Area(A).

Below:
Fig. 4.32 Limiting the spiral tiling to a square unit, shading, and then tiling with the square using the symmetry operation of reflection leads to a visually complex pattern from technically simple beginnings.

Above:
Rochelle Newman and Mar-tha Boles. Budapest Tiling. *1989. Plasticized papers on illustration board. 30" x 30".*

Below:
Fig. 4.33 The dodecahedral net forms a unit for tiling.

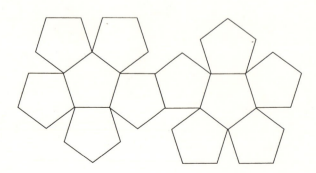

Pentagonal Tilings

Our passion for pentagons and their harmonic properties led us to consider their possibilities in the plane. Although we already knew that a regular pentagon could not be a pro-totile for a monohedral tiling, we were curious about the negative or "left-over" shapes between the pentagons. Since we had previously worked with the pentagonal net for the dodecahe-dron (Fig. 4.33), it visually suggested a starting point for our investigation.

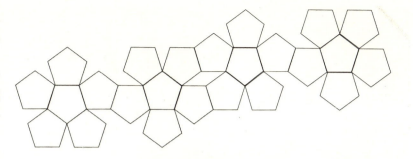

A simple translation of the net as a unit, where the pentagons are joined edge to edge, yields a line group (Fig. 4.34). This group, in turn, may be translated to form a **periodic tiling** if the method of joining edges and vertices is carefully considered. A periodic tiling is one which can duplicate itself when translated in two or more non-parallel directions. The negative shapes can be emphasized by filling in regions thus creating a counterpattern (Fig. 4.35). Filling in areas for contrast is another way to achieve more visual variety. Adding the element of value change or color extends the game almost limitlessly. The appearance of the structure may be changed by judicious addition or deletion of line segments, or by replacing line segments with curves.

Playing with the net suggested a systematic approach to determining the existence of periodic tilings using pentagons. What if other regular polygons were surrounded by regular pentagons? The three diagrams in Fig. 4.36 illustrate possibilities using triangles, squares and hexagons. Each was produced by building out from the original unit, preserving rotational symmetry. An analysis of the overall resulting patterns shows that the method used with the net would also have worked with the square and hexagon, but not with the other where each pentagon is shared by two triangles.

Top to bottom:
Fig. 4.34 Line group created from the "net" unit.

Fig. 4.35   Tiling developed from the line group with negative shapes shaded.

Fig. 4.36   Tilings developed by adding regular pentagons to other regular polygons.

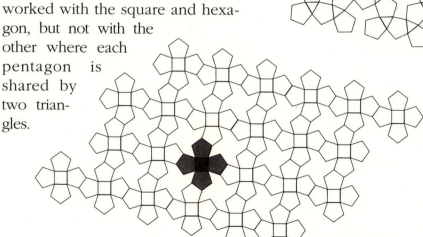

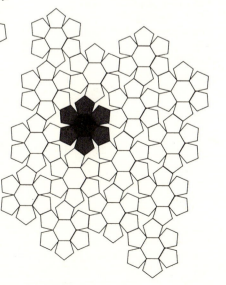

151

Right and below:
Fig. 4.37   Pentagonal line group glide reflected to form a tiling.

Below:
Fig. 4.38   This tiling was created with a unit in which four pentagons surround another. Units were overlapped so that the surrounding pentagons are each shared by two units.

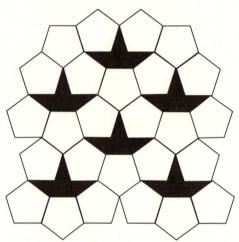

Regular pentagons may be joined in other ways to form periodic tilings. Coupled with the operation of glide reflection, the pentagonal unit forms a line group (LGR), which can then be translated to yield the structure in Fig. 4.37. When four sides of the pentagon, rather than five, are surrounded by other pentagons, a tiling with leaf-like interstices develops (Fig. 4.38).

Embedded within the negative spaces of these tilings are certain recurring angles. Remember that each angle of a regular pentagon measures 108°. Then, when three surround a point, the remaining angle measures 360°-3(108°) or 36°. When two surround a vertex the other angle measure 360°-2(108°) or 144° (Fig. 4.39).

The 36° angle is the vertex angle of the Golden Triangle, so within the pentagonal tilings are golden connections beyond those inherent in the parent polygon. In fact, in the aforementioned tiling, the Lute of Pythagoras lies hidden in the structure (Fig. 4.40).

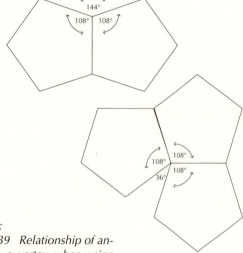

Above:
Fig. 4.39   Relationship of angles to a vertex when using pentagonal prototiles.

Far right:
Fig. 4.40   Lutes of Pythagoras hidden within the tiling of Fig. 4.38.

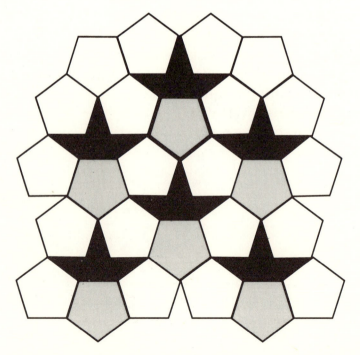

Nonperiodic Tilings

In Fig. 4.41, the rhombic units have angles of 36° and 144°, the resulting form if two Golden Triangles are laid base to base. This rhombus, partnered with another whose angles measure 108° and 72°, plays a major role in the **non-periodic** group of Penrose tilings, named for the mathematician physicist, Roger Penrose, who discovered them and spent a great deal of time investigating their properties. They are particularly fascinating for their numerous golden connections, as well as their relationship to certain natural phenomena.

A question that engaged mathematicians for some time was whether there existed a set of tiles that would tile *only* in a nonperiodic fashion. That is one which would not duplicate itself when translated in two or more non-parallel directions. The first answer to this question was proposed by Robert Berger in 1964, but it required more than 20,000 prototiles. The quest ultimately became to fulfill these requirements with sets containing the fewest prototiles.

In 1974, Penrose found two that worked. These prototiles, which he called *darts* and *kites,* come directly from the regular pentagon and its corresponding pentagram, and, therefore, have Ø connections. The dart is a reflection of the isosceles triangle whose base is Ø and legs are 1. The kite is a reflection of the Golden Triangle with base 1 and legs Ø (Fig. 4.42).

The kites and darts can be fitted together to create a rhombus (Fig. 4.43). Since any quadrilateral will tile the plane periodically, a rule is needed to govern the ways in which the tiles may be joined in order to force nonperiodicy.

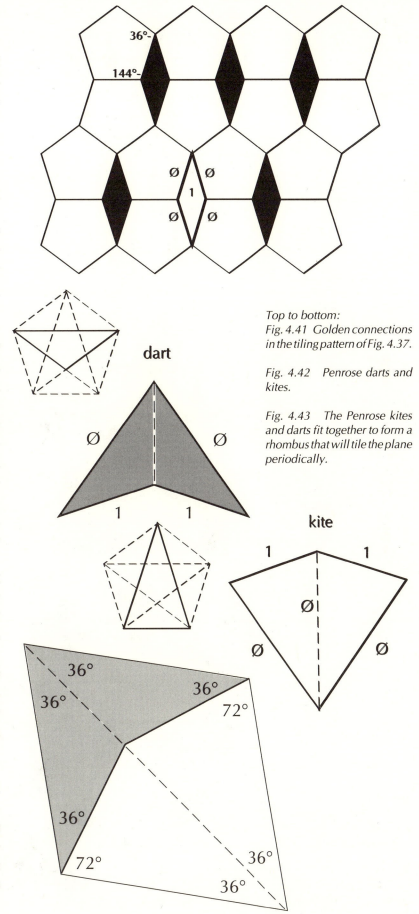

Top to bottom:
Fig. 4.41 Golden connections in the tiling pattern of Fig. 4.37.

Fig. 4.42   Penrose darts and kites.

Fig. 4.43   The Penrose kites and darts fit together to form a rhombus that will tile the plane periodically.

Marking the tiles helps to establish the rule. Draw arcs of two different widths on each tile as illustrated by the heavier and lighter curves in Fig. 4.44. These arcs subdivide the sides in the Golden Ratio. The rule, then, is that tiles may only be fitted together such that arcs of the same width are joined. You can see that this prevents the tiles from forming the previously mentioned rhombus.

Nonperiodic tilings are constructed in a different manner from those we have previously examined. Rather than containing the translational symmetry of the periodic ones, they must grow radially although they need not, and in fact most do not, contain point group symmetry. They can be constructed through a process of trial and error where some choices are dictated by shape constraints. A choice not forced by the particular shape may lead to "dead ends" further on, situations where the spaces to be filled accommodate neither of the protiles nor a combination of them. When that happens, one must back up and reconsider those earlier choices.

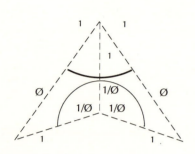
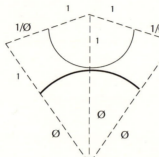

Above:
Fig. 4.44 Arcs that insure nonperiodic tilings with kites and darts.

Below:
Fig. 4.45 A portion of a nonperiodic tiling with kites and darts.

Alas, Escher died before he could know of Penrose's tiles. How he would have reveled in their possibilities.

Martin Gardner
Penrose Tiles to Trapdoor Ciphers

You may be familiar with the work of Dutch artist M. C. Escher, whose intricate tiling patterns with animal and other representational shapes have fascinated artists and mathematicians alike. Many of Escher's tilings began with the simplest of tiling patterns, the regular ones. By carefully cutting a shape from one side of the polygon and adding the same shape to the opposite side, he was able to create periodic tilings of a most unusual kind.

Penrose played with this same technique to create a nonperiodic tiling with prototiles in the shape of chickens developed from the kites and darts (Fig. 4.46). It is interesting to note that in spite of the shape of the tiles, they need only be rotated ,and never reflected, to create an infinite nonperiodic tiling.

In 1979, Penrose discovered another pair of tiles, even simpler than the kites and darts, that will tile nonperiodically. They are a pair of rhombuses, one of which is the one formed by combining a kite and a dart. The other results from joining two Golden Triangles at their bases rather than at a leg as in the kite. The "fat" rhombus has angles of 72° and 108° and the "skinny" one's angles measure 36° and 144°.

The measure of each of the angles

is a multiple of 36°. Therefore, the angles can be named θ (where the Greek letter theta stands for 36°) and 4θ (4x36°) in the skinny rhombus, and 2θ (2x36°) and 3θ (3x36°) in the fat one. Since ten angles of 36° fit around a point in the plane, the possible vertex nets can be determined simply by taking combinations that add up to 10θ. $3\theta,3\theta,4\theta$ or $\theta,\theta,2\theta,\theta,\theta,2\theta,2\theta$ are just two of the many possibilities. However, it turns out that only seven of the possible vertex nets will lead to perfect nonperiodic tilings. These are illustrated in Fig. 4.47.

Upper left:
Fig. 4.46 Penrose's nonperiodic chickens.

Below:
Fig. 4.47 The seven possible vertex nets for perfect Penrose nonperiodic tilings.

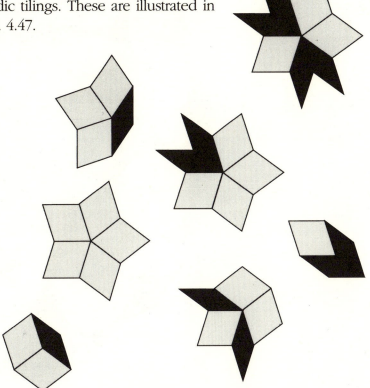

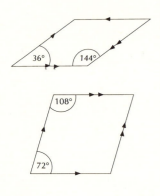

Above
Fig. 4.48 The arrows on the rhombic tiles indicate how the pieces must fit together in order to create a Penrose tiling.

Below:
Fig. 4.49 The characteristics of a Penrose tiling that link it to pentagonal symmetry are the orientation of the decagons and the five sets of parallel lines of tiles.

As with the kites and darts, lines can be drawn on the surface of the tiles to restrict the manner in which they can be joined. Alternately, arrows can be placed on the edges of the tiles where the rule is that tiles may be joined only if arrows coincide (Fig. 4.48). Even with the restrictions, the pattern may break down and fractures appear. When this happens, earlier choices, made when moves were not forced, must be reconsidered.

In a rhombic Penrose tiling there is no true symmetry, but there are some interesting properties related to pentagonal symmetry. Every tiling contains an infinite number of decagons, and each one has precisely the same orientation. Also, if the tilings are shaded in such a way as to color all rhombuses that have sides parallel to

one another, this can be done in five ways (Fig. 4.49). In each case the shaded tiles form sets of implied parallel lines. If we consider all five cases, these lines are consistent with the sides of a regular pentagon.

The nonperiodic tilings discovered by Penrose have some other interesting golden connections. In both the ones formed by kites and darts and those composed of rhombuses, the ratios of the areas of the larger piece to the smaller is Ø. (See Appendix B for a proof of this property.) It is also the case that in an infinite tiling, the ratio of the number of larger pieces to the number of smaller ones is exactly Ø. In fact, one of the questions mathematicians are now asking is whether there exists an infinite nonperiodic tiling that does *not* bear any relationship to the Golden Ratio. Consensus seems to be that there are none, but it has yet to be proved.

The Penrose connection to the natural world can be seen in three dimensions. The elements which compose crystal structures have 3,4, and 6-turn symmetry, and the crystals themselves have translational order. Conventional crystals do not have pentagonal rotational symmetry for the same reasons that pentagons do not tile the plane without gaps. In three dimensions the gaps are known as frustrations and lead to structural instability. However, the discovery of a new group of materials called metallic glasses, or quasicrystals, have structures somewhere between crystalline and amorphous and have a surprising relation to Penrose tiles. Many properties of crystal structures can be investigated by looking at related two-dimensional tilings, but it is the somewhat ordered Penrose tiling with rhombuses that models the somewhat ordered structure of quasicrystals.

Overlapping Pentagonal Tilings

We have looked at pentagons in contiguous relationships. Many possibilities in pure pattern design, however, stem from the overlapping of elements and the interlacing of these overlaps. The design vocabulary of Islamic art relies heavily on these two strategies, and there are many good sources that speak to the particulars of the art of this culture. Our desire is to present a logical approach to design that allows complex structures, similar to Islamic ones, to be built from the knowledge of basic interactions.

Let us, again, take the regular pentagon as the primary unit. When two are overlapped in a regular manner a star icosagon is formed. This simple pairing provides a fertile ground for exploration. The following constructions give three methods for developing tilings based on this unit.

Above:
Eric Kulin. Student work. Units duplicated on the photocopier, manipulated, then colored with pencils and inks.

Below:
Fig. 4.50 Overlapping pentagons form the basis for tiling units related to Islamic patterns.

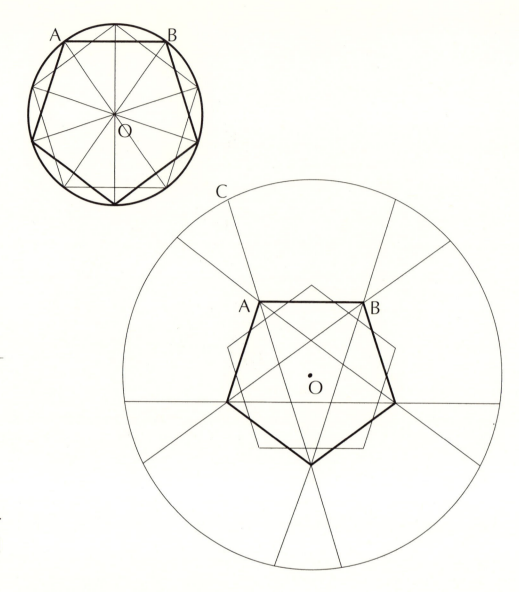

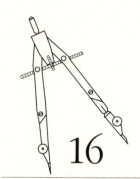

16

Construct a Tiling Unit Based on Pentagons Arranged Edge-to-edge around Overlapping Central Pentagons

Given a regular pentagon within a circle

1. Locate the midpoints of the sides of the pentagon. Draw segments from each vertex through the midpoint of the opposite side to the circle.

2. Connect the points on the circle to create a second pentagon overlapping the given one.

Construct regular pentagons on each side of *one* of the interior pentagons according to the following directions:

3. Erase the circle and the diagonals drawn in Steps 1 and 2, but leave the center, O, clearly marked. Draw all diagonals of *one* of the pentagons. You will see a pentagram.

4. Extend the diagonal through A such that AC = AB.

5. Draw a circle with radius OC.

6. Extend each of the sides of the pentagram, in both directions, to meet the circle. There will be two extensions at each vertex.

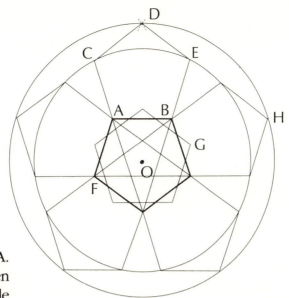

7. Open the compass to measure CA. With the metal tip on first C and then E, cut two arcs that intersect outside the circle. Label the point of intersection D.

8. Draw a circle with radius OD. This will determine the placement for the outermost vertices of the ring of pentagons.

9. The vertices can be located by laying a straightedge along two opposite vertices of the overlapping central pentagons. For example, locate H by placing a straightedge along F and G.

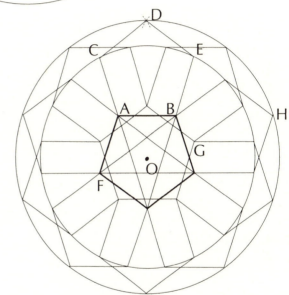

10. Repeat Steps 6 and 9 to obtain a second ring of pentagons adjacent to the other inner pentagon.

11. Erase extraneous line segments.

Now the result is a tiling unit with overlapping pentagons.

> *The second ring of pentagons could be gotten by taking a tracing of the first group and rotating it to the appropriate position for the second group.*

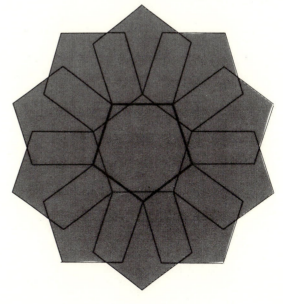

Before tiling, the skeletal structure may be embellished if desired. Perhaps you will want to experiment with overlapping two units. Once you have decided on your unit, translate it to form a line group, and then translate the line group to create a tiling (Fig. 4.51). Part of the fun is to try different translations. Or, just have a good time playing with color or shading in the tiling on the opposite page!

Fig. 4.51 These possibilities use simple interlacing techniques.

Studio tip: Using a computer or photocopying onto acetate or tracing paper will save several eons of frustration when trying to tile the plane.

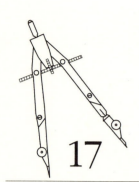

17

Construct a Tiling Unit Based on Pentagons Arranged Vertex-to-vertex around Overlapping Central Pentagons

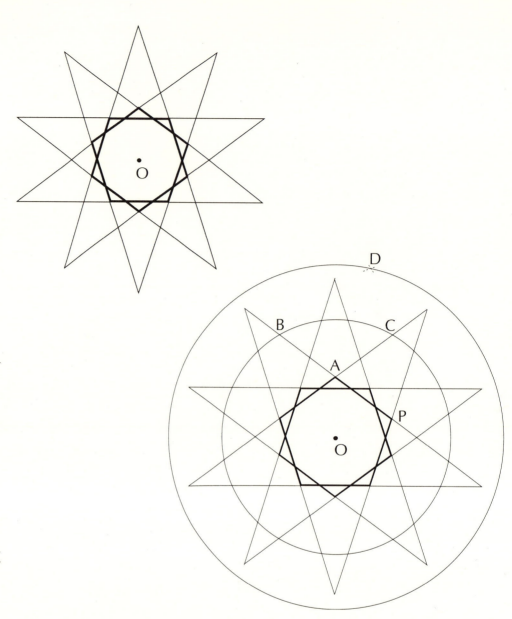

Given a pair of overlapping regular pentagons with center O

(See previous construction)

1. Extend all the sides of the pentagons to form a star icosagon.

2. Locate B so that $\overline{AB}$ is congruent to a side of one of the original pentagons (such as $\overline{AP}$). Draw a circle with radius OB.

3. Open the compass to measure BC. With the metal tip on B cut an arc in the exterior of the icosagon.

Set the compass to measure CA and with the metal tip on C cut another arc to intersect the one just drawn. Label the point of intersection D and draw a circle with radius OD.

These two circles and the original figure will now allow the construction to be completed by drawing line segments through existing points.

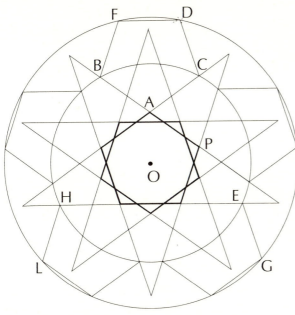

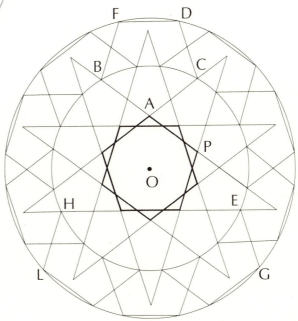

4. Lay a straightedge along C and E and draw $\overline{DC}$ and $\overline{EG}$. Repeat along B and H to draw $\overline{BF}$ and $\overline{HL}$. Continue through other points of intersection of the pentagram and circle to obtain the sides of a ring of five small pentagons.

5. Draw the outer sides of the pentagons by connecting adjacent points on the larger circle, such as D and F.

6. Repeat the process using the other pentagram to obtain the overlapping ring of pentagons.

7. Erase extraneous line segments and circles.

Now the result is a tiling unit with overlapping pentagons.

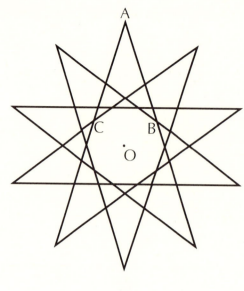

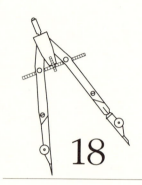

18

Construct a Tiling Unit with an Outer Ring of Ten Pentagons

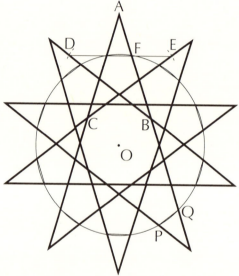

Given a pair of overlapping pentagrams that form a star icosagon centered at O

*(See Step 1 of the previous construc-
tion)*

1. Open the compass to measure AB.
With the metal tip on B, cut an arc to
locate point D. Without changing the
setting, place the metal tip on C and
cut an arc to locate point E.

2. Draw DE so that a small pentagram
is formed.

3. Draw a circle with radius OF to find
the points on each of the other nine
arms of the pentagrams that determine the outermost sides of the ring of
small pentagons.

4. Connect the points of intersection of
the arms of the pentagrams and the
circle to form the outermost sides of
the small pentagons ($\overline{PQ}$ for instance).

5. Erase extraneous line segments and
the circle.

Now the figure is a tiling unit consisting of a ring of ten pentagons.

*One star icosagon appears
within the ring of pentagons.
Another becomes evident with
the inclusion of the Penrose
kites which are nested among
the pentagons.*

Above:
Juliet Burley-Mason. Student work Pen and ink detail from a project that used the concept of tiling.

Construction 18 gives a unit which actually does appear as a structural foundation for many Islamic tilings. With the inclusion of selected diagonals various interior figures can be formed. Further embellishment provides variety and surprise. We encourage you to do a more in-depth study of Islamic tilings using the many good resources available.

We know that tilings can fill the plane infinitely if there is no stopping point. In our discussion of a few of the infinitely many tilings, we choose to stop here. We hope our stopping place is just a starting point for you.

Problems

Above:
On the computer tiling patterns take on a more artful appearance when coupled with layering and shading.

|4| Create a dihedral tiling using a pair of Dynamic or Ø-Family Rectangles.

|5| Choose a regular n-gon. By fitting regular pentagons to no more than n-1 of the sides, create a unit that can be tiled using any method described in this chapter.

|6| a. Use a subdivision of the prototile in one of the regular tilings to create a tiling that has prototiles of irregular shapes.

b. Use a subdivision of the prototile in a non-regular monohedral tiling in order to create a dihedral or trihedral tiling.

|7| Choose any rectangular or triangular grid. Using key points (your choice) superimpose a net of overlapping circles to form a tiling.

|8| Use the templates in Appendix E to make a set of Penrose rhombuses. You will need about 16 fat ones for every 10 skinny ones. Tile a 10" x 10" square without any fractures.

|9| Construct one of the units given in Constructions 16, 17 or 18 and alter the interior to create a different looking unit.

|1| By using the chart on page 136 and the formula, find one of the six remaining vertex nets and give its order.

|2| Construct a dihedral tiling, different from any illustrated in the text, using two triangular prototiles.

|3| Using a minimum of ten tiles, construct a tiling that has a nonregular hexagonal prototile.

Projects

1 Create a fantasy bit of flora or fauna within a tile unit(s). Using any of the principles of tiling discussed in this chapter, cover a portion of the plane. Emphasize the natural element as opposed to the underlying tile structure. Be sure to make it beautiful.

2 Design an asymmetric motif within a square. Using the operations of translation, rotation, reflection or a combination of these, create a tiling. Enhance in any way in order to make it lovely.

3 Develop a game, based on Chess or Checkers, that utilizes a board with hexagonal units. Include written rules and descriptions (drawings, sculptures, diagrams) of playing pieces.

4 Choose a square, triangular or hexagonal prototile. Subdivide it harmonically and tile

a portion of the plane. Alter the tiling with color, value, or black and white in such a way that a new tiling is formed and the original one is hidden.

Below:
Claire Melanson. Student work. Combination of hand drawing and computer tiling.

5 | Develop a set of children's blocks that utilizes the ideas of plane group symmetry as related to tilings.

6 | Transform a polygonal tiling into a non-polygonal one by replacing line segments with arcs. Use materials of your choice and a maximum of four colors to finish it.

7 | On heavy paper, develop a geometric tiling that combines straight lines and curves. Color as you wish. Turn the piece into a jigsaw puzzle either by superimposing a commercial structure on the back, or developing your own. Cut into pieces and drive your friends crazy by having them put it together.

8 | Research the tiling work of Dutch artist M. C. Escher. Create an Escher-like tiling.

9 | Use any of the problems as the basis for an artwork. Be concerned with texture, color, craft and content.

Below:
The concepts of tiling and circles merge in this pavement pattern.

Further Reading

Bezuska, Stanley and Margaret Kenney and Linda Silvey. *Tessellations: the Geometry of Pattern*. Palo Alto: Creative Publications, 1977.

Bourgoin, J. *Arabic Geometrical Pattern and Design*. New York: Dover Publications, Inc., 1973.

Coxeter, HSM. and M. Emmer, R. Penrose and M. L. Teuber. *M. C. Escher: Art and Science*. Amsterdam: North Holland, 1986.

Critchlow, Keith. *Islamic Patterns*. New York: Schocken Books, 1976.

Gillian, Edmund V. *Geometric Design and Ornament*. New York: Dover Publications, Inc.,, 1969.

Grunbaum and Shephard. *Tilings and Patterns*. New York: W. H. Freeman and Co., 1987.

Peterson, Ivars. *Islands of Truth*. New York:W. H. Freeman and Co., 1990.

Schattschneider, Doris. *Visions of Symmetry: M. C. Escher*. New York:W. H. Freeman and Co., 1990.

Wade, David. *Geometric Patterns and Borders*. New York: Van Nostrand Reinhold, 1982.

Wong, Wucius. *Principles of Two-Dimensional Design*. New York: Van Nostrand Reinhold, 1972.

5 Fractals

> *Everything in Nature is capable of mathematical expression if the conditions are only sufficiently well known and the mathematics are sufficiently complicated. The real difficulty is to select a fundamental conception which will admit of modification when new factors are introduced.*
>
> Theodore Andrea Cook
> The Curves of Life

Artists have struggled with depicting their perceptions of the world ever since cave people first put hand prints to walls. For centuries, mathematicians, too, have labored to develop models of the way things work that extract the essence of their perceptions. In the process of objectifying ideas, the individual first understands for him or herself. Then he or she develops a structure which becomes the vehicle for communication with others. The painter chooses a surface, defines its boundaries, and proceeds to put down marks. The mathematician also "frames a space" by setting the boundaries that allow him or her to define a problem, and then proceeds to put down marks. Both have a common desire to connect the world "out there" to the world "in here." It is a desire to order, categorize, simplify, and clarify; a characteristic of many human endeavors.

Mathematicians and artists are passionate about different aspects of reality, and the models they use depend on the problems they choose to solve. The function of a painting is different from that of a mathematical model, just as a paintbrush differs from an equation. Despite the differences, the overriding goal is the same: to heighten awareness of the world. And yet each model is only an approximation of a small part of reality, never able to be substituted for any portion of it.

The complexity of our natural world has not changed since the time of Euclid. However, the types of problems humans confront have changed enormously. Euclid never had to face a traffic jam on the freeway as he drove to work on Monday morning. He was not bothered by static in his telephone nor on his television set, and there was never a large oil spill off the coast of Greece. Unfortunately, these are commonplace problems of the modern world. But it is only recently that we have had the ability to create mathematical models that describe them and aid in their solutions.

Natural forms and systems defy easy explanations. They have quirks; they have irregularities; they have twists. Obviously, Euclid could see the differences between geometric abstractions and natural forms, but these phenomena did not suggest problems to him. The task he set for himself was to organize into a logical, coherent body, with tight limits, all the geometry known at that time. Evidence indicates that the "nothingness" of space was a difficult concept for the Greeks, so that Euclid's mathematics is one descriptive of stationary, invariant, simple forms. The factor of time never entered into his

descriptions. He had a powerful sense of the abstract which allowed him to describe essential qualities and to see common physical properties in those forms. It is a tribute to Euclid's insight that his textbook, *Elements,* was used for over 2,000 years and is still read today.

Top:
The eye of the camera visually models the natural world.

Center:
J. V. Ruisdael. A Distant View of Haarlem from the Northwest. *Rijksmuseum, Amsterdam, the Netherlands.*

43x38 cm. (A ratio of 1.13:1) Painted in the early 1670's, this landscape typifies Ruisdael's passion for accurate depiction of the clouds, trees, and gray skies he observed in the world around him. He was the most famous of the Dutch landscape artists.

Bottom:
Vincent Van Gogh. A Cornfield, with Cypresses, 1889. *Reproduced by courtesy of the Trustees, The National Gallery, London.*

28.5x36" (A ratio of 1:1.26 — roughly a $\sqrt{\emptyset}$ Rectangle). It is the artist's empathy with the natural world that pulses through agitated, exultant brushstrokes and vibrating color. It is a heart bursting with love and compassion for life itself: raw feelings controlled by composition; the intellect at the service of emotion.

172

In the intervening centuries, as new problems and new technologies came into being, there were persons who met the challenges with other creative solutions. Scientists in the areas of chemistry, biology, geology and meteorology have been looking at situations involving unpredictable conditions for quite some time. They turned to the mathematical community to provide the models that would offer possible solutions.

How could one describe a cloud formation or provide a method for duplicating one? Artists have been sharing their perceptions of these for centuries. It is only now, in the Twentieth Century, that scientists have been able, with the use of the computer, to visually model what artists have previously defined through the use of pencils and paintbrushes.

Above:
Paul Cezanne. Mont Sainte-Victoire Seen from Bibemus Quarry. ca. 1897. The Baltimore Museum of Art: The Cone Collection, formed by Dr. Claribel Cone and Miss Etta Cone of Baltimore, Maryland. BMA 1950.196.

23.5x32" (A ratio of 1:1.25 — roughly a $\sqrt{\emptyset}$ Rectangle). Cezanne's interest in a painting was as a self-contained structure; a "construction after nature". While expressing emotions through pure form and color, he never abandoned direct observation, always seeking to "see in nature the cylinder, the sphere, and the cone".

Left:
Benoit Mandelbrot. Planet Rise. Computer image. He states, "Clouds are not spheres, mountains are not cones, coastlines are not circles, and bark is not smooth, nor does lightening travel in a straight line".

Cezanne was as interested in the structure of a painting as Mandelbrot is interested in the structure of a coastline. It is the form in which their passions are expressed that differs.

Clouds are forms that appear similar over time, but never are exactly the same. They are just one aspect of the larger system which is weather. People talk about the weather, complain about it, and then blame the forecaster for inaccurate predictions. Somehow, it is felt that he or she, as a person of science, should be able to be more accurate. But weather, like many other systems, is composed of a multiplicity of details, each one affecting the others in subtle and unpredictable ways. Weather is governed by the same laws as those that rule planetary motion, and when these motions are described they are done so through approximations. Science is based on approximations — in the input and approximations in the outcome. The world, however, does not truly operate by approximating. It simply *is*.

What scientists have discovered, based on the work of meteorologist/mathematician Edward Lorenz, is that everything is affected by "sensitive dependence on initial conditions", what has come to be called the butterfly effect. That is, minute changes at the beginning of a process, even a motion as gentle as the beating of a butterfly's wings, may result in tremendous variation later on. Therefore, approximations cannot account for all of the myriad possibilities.

At the turn of the century, some mathematicians were happening upon certain geometric figures that had aberrant properties. These they labelled *monsters*, and considered them interesting, but useless. What did these *monster curves* have in common with such natural phenomena as weather? There was a connection. But it required a person who saw problems, not disciplines; one who was not locked into a particular frame of reference, to see that these could serve as models for describing the intricacies of the natural world. That person was Benoit Mandelbrot, born in 1924, a maverick mathematician, who thought in terms of shapes and trusted his visual intuition.

Since 1952, he had been looking at such diverse phenomena as the flood patterns of the Nile, cotton prices in the United States, trends in the stock market, and noise in communication systems. He saw connections where others saw confusion; he saw patterns where others saw randomness. He found the mathematics to articulate the problems. He had the vision and had access to the one tool that would allow him to examine the problems in detail—the computer.

At first, he had intuitions but no name for them. Then, in 1975, he coined the word **fractal**, derived from the Latin verb *frangere*, to break, and the adjective *fractus*. A new geometry was in the making and given life by its naming. It was descriptive of conditions that change over time, and forms too complicated to fall within Euclid's framework. It is a reminder that mathematics is a living language, not a dead, forgotten tongue.

The computer is the prime tool for doing the repetitions required to see fractal growth. Without it, fractal geometry, very likely, would never have come into being. At this juncture,

the skills and sensibilities of both mathematicians and artists overlap. Mandelbrot's work to capture the physical essence of nature allows the computer screen to act as a canvas. Those artists who find the computer a compelling tool now have greater flexibility in building forms that more nearly capture the visual essence of

the world, especially in terms of its textural qualities. It has also enabled lay persons to see what the experts are privy to. But in this chapter we will work primarily with tools and a scale that the hand and eye can manage. We are interested in introducing the basic concepts of fractals without repeating what is done so well in the books listed in the Further Reading at the end of the chapter. We hope our work here stimulates interest for those new to these ideas.

Above:
Benoit Mandelbrot, father of fractal geometry.

These two pages:
Fig. 5.1   In each diagram the actual distance from A to B remains constant. The dotted line shows the previous night's path, and the solid line indicates the current route.

On Sunday Decat zips home.

A ─────────────────────────── B

Sunday

What is Fractal Geometry?

Recall our little friend, Decat, from the Symmetry chapter. He is hurrying home after visiting his girlfriend, Phiphi, on a hot summer Sunday. She has given him a gallon of his favorite ice cream, chocolate macadamia nut crunch, to put in his freezer. He is pedalling as fast as he can down the street, following the straight and narrow path between their two houses. On Monday he might be more interested in the road less travelled. If he is walking his sweetheart home in the moonlight, he will look for ways to lengthen the walk, for there is no advantage to getting from his house to hers along a straight line.

Fig. 5.1 demonstrates a rule for the romantic couple's walks. Remember that the process is more theoretical than practical and does not take actual distances into consideration.

On Monday, as they leave his house, point A, they turn 45° to the right, then, at the halfway point, 90° to the left toward her house, point B, so that their route becomes the legs of an isosceles right triangle. That did not leave much time for poetry recitation.

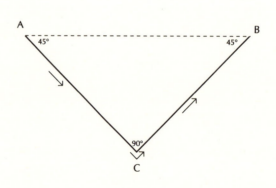

Monday

On Tuesday night they repeat the same process, 45° right, 90° left, between the first two points, A and C, of Monday's path and then 45° left, 90° right between the second two points, C and B.

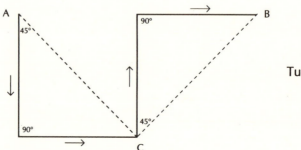

Tuesday

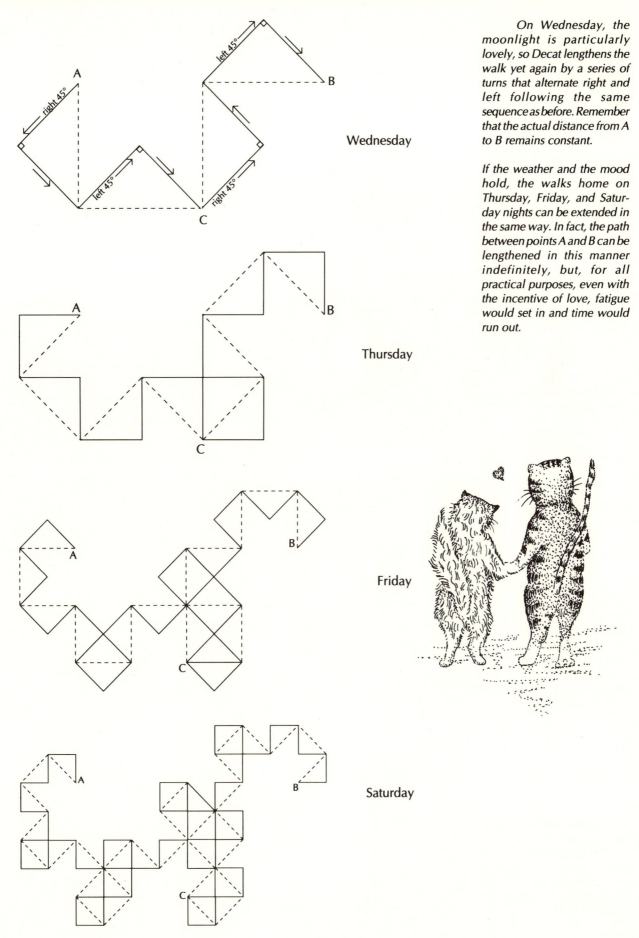

Wednesday

Thursday

Friday

Saturday

On Wednesday, the moonlight is particularly lovely, so Decat lengthens the walk yet again by a series of turns that alternate right and left following the same sequence as before. Remember that the actual distance from A to B remains constant.

If the weather and the mood hold, the walks home on Thursday, Friday, and Saturday nights can be extended in the same way. In fact, the path between points A and B can be lengthened in this manner indefinitely, but, for all practical purposes, even with the incentive of love, fatigue would set in and time would run out.

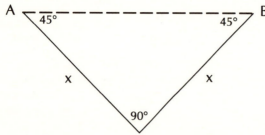

A 45° ———————— 45° B

x   x

90°

Above:
Fig. 5.2 Monday's route forms an isosceles right triangle.

we substitute a particular numerical value of *x*. Now we can check the length of the walk each evening. Every time we evaluate the function, we replace the original *x* with the previous night's distance.

Just how long is their walk? Let us suppose that it is one mile between the two houses. We will label each part of Monday's route **x** (Fig. 5.2). Since each part represents one leg of an isosceles right triangle, the lengths will be the same. By the Pythagorean Theorem $(a^2 + b^2 = c^2)$, then:

$$x^2 + x^2 = 1,$$
$$\text{so } 2x^2 = 1,$$
$$\text{or } x^2 = \frac{1}{2},$$
$$\text{and } x = \frac{1}{\sqrt{2}}.$$

Since x represents only one half the distance from A to B, the actual length of the walk is 2x, or $\frac{2}{\sqrt{2}} = \sqrt{2} \approx 1.4$, or almost one and one half times the actual distance from A to B. We could use similar reasoning to show that each day the new path is 1.4 times that of the previous evening. When we devised a rule for the walk, it was a purely geometric one. Now that we know the rate of increase each day, we can rephrase the rule in algebraic symbols: f(x)=1.4x, read *f of x is 1.4 times x*. f(x) means the **function** of *x* and is the "answer" we get when

Distances are given to the nearest tenth of a mile.

f(x) = 1.4x

Starting with x=1 (Sunday's distance), we get:
f(1) = (1.4)(1) = 1.4   Monday
f(1.4) = (1.4)(1.4) = 2   Tuesday
f(2) = (1.4)(2) = 2.8 Wednesday
f(2.8) = (1.4)(2.8) = 3.9   Thursday
f(3.9) = (1.4)(3.9) = 5.5   Friday
f(5.5) = (1.4)(5.5) = 7.7   Saturday

With a walk that is 7.7 miles on Saturday night, Decat has plenty of time to work up his courage to propose. He does!

What if the path doubled in length each night? We could make a rule that says to turn right 60° then left 60° to form sides of an equilateral triangle between each two previously existing points (Fig. 5.3). Now the algebraic rule is the function f(x) = 2x, and the nightly distances would be as follows:

f(x) = 2x, and we begin with x=1.
f(1) = 2(1) = 2
f(2) = 2(2) = 4
f(4) = 2(4) = 8
f(8) = 2(8) = 16
f(16) = 2(16) = 32
f(32) = 2(32) = 64 miles on Saturday night.

If Decat and his love walk at about two miles per hour, this "Saturday night walk" would last into Monday!

Notice in this example the difference between 1.4 and 2.0 seems, for all practical purposes, to be relatively slight — only .6. Yet, in a very short time, the discrepancy gets magnified, so that over many repetitions that .6 difference becomes huge (Fig. 5.4). You may have already noticed this phenomenon when doing constructions with compass and straightedge. Even if the directions are followed precisely, minute variations, such as the thickness of a pencil line, can distort the end result. A business could gain or lose a fortune, depending on whether, in the course of a year, prices are rounded up or down to the nearest penny. Again, the outcome is dependent on minute variations at the beginning.

Since natural phenomena are particularly sensitive to initial conditions, it is foolhardy, and potentially dangerous, to assume that long range predictions can be based on short run data. In this century, especially, we need to be aware of the subtleties of relationships and minute variations. Although we have set artificial boundaries between cities, countries, and continents, natural events are not constrained by those boundaries. Consequences of decisions made in regard to these systems do not necessarily last for only a moment or a single generation, but rather reverberate through space and time. Therefore, acts in our own backyards can have global consequences for generations to come.

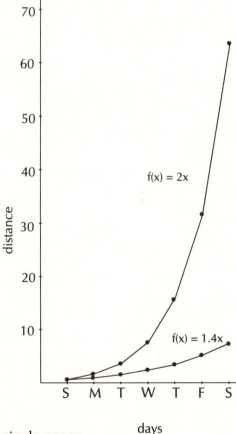

Left:
Fig. 5.3   If the turns are each 60°, an equilateral triangle, rather than an isosceles right triangle, describes the path on Monday night.

Below:
Fig. 5.4   The graph illustrates how quickly the distances diverge using the two different functions.

Below:

Fig. 5.5 Two variations on the configuration of the path of Decat's walk. Each of the figures shows what the walk might be after ten nights with a random left-right sequence of turns.

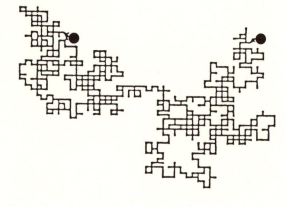

Below:

Fig. 5.6 This figure, iterated on the Macintosh computer, illustrates the 50,000th generation of Decat's walk. It is commonly known as the Dragon Curve. Described by an equation one line long, brevity and simplicity disguise its richness.

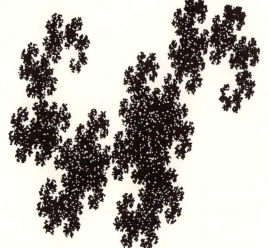

There are other ways Decat could have lengthened the walk in a regular fashion. He alternated the 45° turns, right, left, right, left, etc. (R L R L ...). He might have chosen to turn left first in a L R L R... pattern. He might have decided always to turn in the same direction; R R R R ... or L L L L .... Or, letting chance play a role, he could have stopped and flipped a coin each time he needed to make a turn; heads, turn right, or tails, turn left. Under the latter circumstances, the walk would become less patterned, more random (Fig. 5.5).

Through this exercise some of the essential characteristics of fractal geometry have surfaced. One is the idea of a repetitive process, called **iteration**, usually described by a simple rule. The result in one step becomes the initial information in the next. It is the number of iterations of the rule that determines the complexity of the final form. The configuration of Decat's path on a particular day is determined by the number of times the iteration has been performed and is known in most texts as the **order** of the curve. We have chosen to substitute the word **generation** for order, because, for us, it more clearly describes a specific place in a continuum. For example, the straight line on Sunday is a first generation curve and the very wiggly one on Saturday is the seventh generation. It is evident that the computer, which never gets angry, bored, frustrated, lazy, or just plain tired, is the best tool for producing images of later generations (Fig. 5.6).

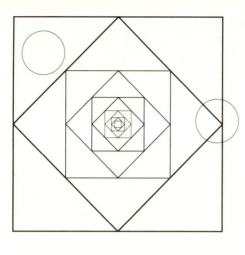

Left and center:
Fig. 5. 7   Each of the two figures displays the property of similarity and hints at the notion of self similarity because of the scaling. However, if the only visible parts were the circled sections, one would have no clue as to the character of the entire figure.

Below:
Fig. 5.8   Notice that half of this curve can be rotated through an angle of 90° to produce the other half. This is true for all generations of this curve.

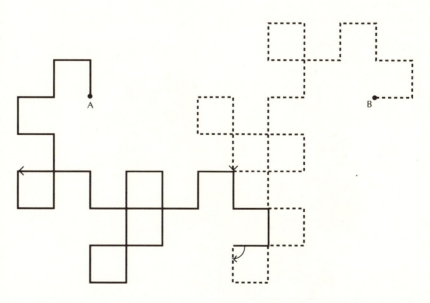

The major property of fractals, however, is the quality of **self similarity**. The notion of similarity is not new to you. It is embedded in the Divine Proportion which gives rise to the Golden Ratio; in the concept of the reciprocal of a rectangle; and visibly evident in logarithmic and Archimedean spirals. The property of self similarity, though, is even more restrictive than that of similarity, in that the shape of the figure remains constant regardless of the level of magnification, or scale. If we look at Decat's path on Saturday night (7th generation) we see that the dotted portion (Fig. 5.8) represents a 6th generation curve, two 5th generation curves, etc. Taken to a very large number of iterations, any portion of the figure contains the character of the entire curve.

Scale invariance is the term used to describe this phenonemon. If we observe a twig, for example, the branching pattern is the same as that found in the whole tree. Whether we look at a part or the whole, the pattern structure is the same. Changing the distance from which we observe, zooming in and out, reveals structures that are alike in kind if not in detail.

Meanwhile, back in our story, De-cat and his new blushing bride, Phiphi, decide to honeymoon at a seaside resort. (Obviously the poetry was effective!) As they check the map for the coordinates of their destination, they notice that the length of the coastline appears very short. However, once on the beach, the couple finds the walk along the shore to be much more than it appeared on the map. And to the ants stealing crumbs from their picnic, even short distances seem infinite. At each of these levels the intrinsic character of the coastline does not change, but with increasing magnification the curve lengthens. It cannot follow the perfect mathematical model for self similarity, yet remains alike in kind, if not in detail, regardless of the size of the viewer. Mathematically we can conceive of things getting smaller to infinity, but at some level the natural coastline will be reduced to a collection of atoms and subatomic particles, bounded by the finiteness of the physical world.

How can the differences in the walks between the two houses, or between two points on the shoreline, be described? How can we label the deviation from the straight and narrow? Both occurred along a nonlinear path. Each fills more of the plane than does a line, yet neither totally covers it.

In Euclidean geometry, a line is a one dimensional figure and the plane has a dimension of two. Therefore, the curves discussed have a dimension somewhere between 1 and 2, and, thus, must be represented by a fraction or an irrational number. Fractal dimension is the term Mandelbrot coined to describe the wiggliness of a fractal curve. For example, the dimension of the Dragon curve is greater than the dimension of the coastline, since the former touches its own path many more times than the latter, thus covering more of the plane.

Generating Fractals

One way to develop a fractal image is to begin with a figure and an **algorithm** (a mathematical rule). The figure is called the **initiator**, and the second generation curve (the resulting figure after one application of the algorithm) is call the **generator**. Like DNA, the generator carries the "genetic code" for all future generations of the fractal image. The algorithm describes what must be done to the figure to produce each subsequent generation. In Decat's walk, the initiator was a line segment (the straight path between the two houses), and the algorithm required that the segment be replaced by the legs of an isosceles triangle in a particular right, left sequence. Built into the process is iteration, using the results of one

generation for the initial condition of the next. This suggests a loop, and, in fact, when generated on a computer, the process of looping allows for the creation of a very complex image coming from a few simple statements and little storage capacity.

Line Segments as Initiators

Given your garden variety line segment, what can be done to transform it? Using symmetry operations is one way to allow for developing potentially more interesting generators for the fractal process. First, consider the flexibility of dividing the segment into thirds and then manipulating these in a variety of ways. One operation is to bump out one or more sections and replace it or them with two line segments suggesting sides of a triangle. Fig. 5.9 shows a variety of ways in which this can be done.

Below:
Fig. 5.9 Generators developed using the concept of replacing thirds of the segment with sides of equilateral or right triangles. To produce the next generation, replace each line segment in this generation with a scaled copy of the generator.

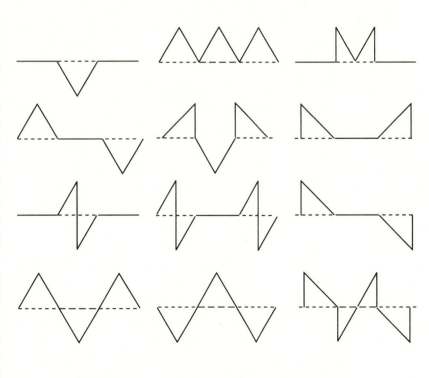

Right:
Fig. 5.10   Cantor dust is created by dividing a line segment into thirds and removing the middle third.

Below:
Rochelle Newman. Autumn Fractal. *Acrylic paint on water-color paper with woven elements.*

Repetition and restraint lead
an artist to greater freedom
and creativity.
            Bridget Riley
20th Century English Artist

There are infinitely many other ways in which a line segment can be manipulated. There is nothing sacred about using the middle third. More complex generators may be obtained by subdividing the initiator into more than three segments or by using forms

other than triangular ones. The "bumped" portions may suggest the sides of any polygonal figure, but curves are not allowed. Each segment of the generator is replaced with a scaled copy of itself to produce the next generation, and there is no way to replace a curved portion with such a copy.

A question for exploration: "Is there an advantage to more complexity?" The answer hinges on the original intent of the person generating the fractal image. Is the concern for aesthetics? Is the concern for relating to the natural world? Is the concern for mathematical elegance? Ultimately, these concerns may be totally interrelated. We encourage you to examine them in light of your own investigation of fractal images.

In the previous examples, the generator was created by making the line segment longer and more complex. If the algorithm requires that the segment be divided into thirds, and the middle third removed, a different kind of fractal can be generated. The greater the number of iterations, the more dustlike the appearance. This fractal, called *Cantor dust*, was named after Nineteenth Century, German mathematician, Georg Cantor (Fig. 5.10).

When creating fractals, the important thing to remember is to be consistent when applying the rule. This is made easier if the fractal is generated on a computer, since the rule is incorporated into a "loop". An artist does not always want to play by the rules, but may want to incorporate fractal concepts into his or her work. One of the advantages to working by hand is that the designer is free to decide when to be consistent and when to deviate.

Branching

Consider the same line segment, but change the algorithm to one that determines a branching pattern. Suppose we require, at one endpoint, a branching of 90° where the branches are 1/2 as long as the original segment. This could happen in a number of ways, as illustrated in Fig. 5.11. Therefore, we need to add a placement clause to the algorithm. Imagine an extension of the segment at the end to be branched. Then let half the 90° fall on either side of this axis. This will insure that the branching is not "lopsided".

We can simplify the quantity of words needed for such an algorithm by using mathematical symbols. Let r stand for the length of each new branch as compared to the previous line segment and assume the initiator to have a length of 1. Then let θ designate half the angle of branching.

Let $r = 1/2$. The subsequent lengths of branches could also be written as an iterative function, $f(x) = \frac{1}{2}x$, where x is the length of the segment in any generation.

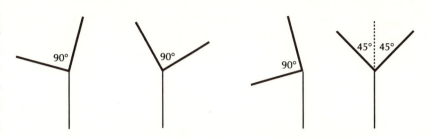

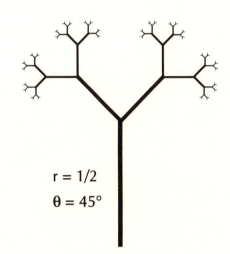

$r = 1/2$

$\theta = 45°$

Above:
Fig. 5.11 Four of the infinitely many ways a 90° branch can be added to the end of a line segment.

Left and below:
Fig. 5.12 Two branching patterns developed from different values of r and θ.

Beginning with x=1 we have:

$$f(1) = \frac{1}{2}(1) = \frac{1}{2}.$$

$$f\left(\frac{1}{2}\right) = \frac{1}{2}\left(\frac{1}{2}\right) = \frac{1}{4}.$$

$$f\left(\frac{1}{4}\right) = \frac{1}{2}\left(\frac{1}{4}\right) = \frac{1}{8}.$$

$$f\left(\frac{1}{8}\right) = \frac{1}{2}\left(\frac{1}{8}\right) = \frac{1}{16}.$$

and so on.

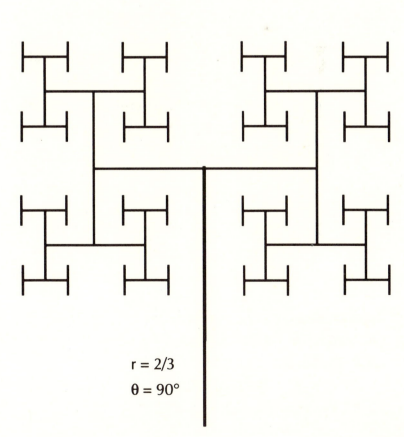

$r = 2/3$

$\theta = 90°$

Above:
Natural branchings are not so regular due to the effects of a multiplicity of outside forces.

> *In terms of the body's resources, blood is expensive and space is at a premium. The fractal structure nature has devised works so efficiently that in most tissue, no cell is ever more than three or four cells away from a blood vessel. Yet the vessels and blood take up little space, no more than about five percent of the body.*
>
> *James Gleick*
> *Chaos: The Making of a New Science*

A Design Problem

Let us consider the design of a container. Packing a simple solid with static items is a fairly easy problem to solve, despite the billions of dollars spent by industry on just such endeavors. Any grocery shelf offers a variety of examples of basic solutions to problems of packaging, ranging from bags to bottles to boxes to cans.

But, what if a container needs to be packed with a multitude of contents all of which are dynamic and in a constant state of growth and decay? Additionally, the package must grow and sustain itself. Living organisms are just such containers. Trees, lungs, circulatory systems all face the same problem; how to quickly get the maximum amount of a precious material from the source to the furthest points in the system. Nature's fractal branching patterns allow for squeezing large surface areas into limited space, in concert with all the other systems at work in a particular organism.

Polygonal Fractals

This time, let the initiator be a polygon rather than a line segment. We can devise a rule that either subdivides the sides of the polygon, in which case the line segments act as they did in the previous examples, or we can devise a rule that breaks up the interior of the polygon in a systematic fashion. Fig. 5.13 demonstrates three ways that equilateral triangles can be used as initiators for fractal images.

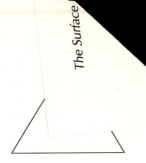

initiator - first generation

This fractal is the Koch snowflake curve

This page:
Fig. 5.13   In each case the initiator is an equilateral triangle. The second (gen-erator), third and fourth generations are shown. In the first two columns, the sides of the triangle are divided into thirds and the middle third replaced with the sides of another equilateral triangle. In the third case, the interior of the triangle is subdivided into four triangles and the middle one is dropped out.

This page:
Fig. 5.14 Six generations of two different fractals with square initiators.

initiator
first generation

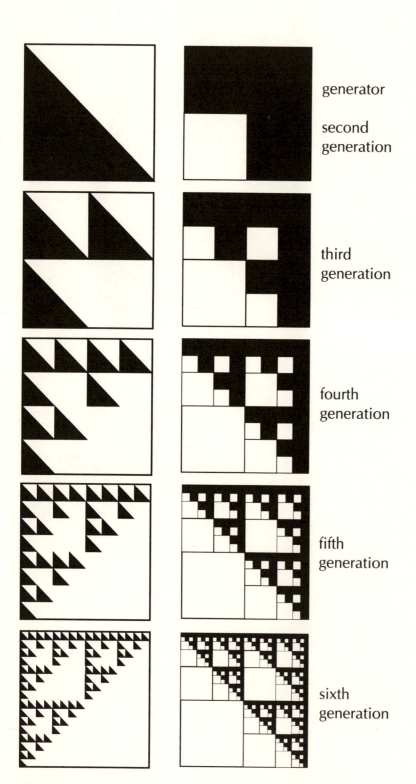

generator

second
generation

third
generation

fourth
generation

fifth
generation

sixth
generation

Fig. 5.14 shows two different fractal patterns generated from a square initiator. In the right-hand column, the algorithm requires the square to be subdivided into four squares and the lower left one removed, leaving the linear outline. In subsequent generations, the process is repeated, covering only the three remaining squares within each square with copies of the generator. The same rule could be followed eliminating the linear components and the resulting pattern would be less visually complex.

In the column on the left, the algorithm dictates that the initiator be subdivided into two triangles by drawing in the diagonal. In this case, the pattern looks to be visually logical, but a slightly different twist is given to the rule. Although the generator appears to be divided into only triangular regions, the rule uses both square and triangular subdivisions. The third generation shows that copies of the generator cover the upper squares and the lower left one, and that the lower right is eliminated. The full algorithm is not visually evident until the third generation.

Above:
Fig. 5.15   A fractal quilt.

Despite the fact that elaborate fractal structures are developed by using the computer, even the first few generations offer a creator new melodies to orchestrate in harmony with such concepts as symmetry and tilings. By using fractal geometry to break up the interior of a polygon or add to its exterior, an infinite variety of patterns can be achieved.

In Fig. 5.15, the second through sixth generations of one of the previously illustrated fractal patterns form a horizontal strip that was rotated to create a unit. The unit was translated once and then half of the unit was used again. One's aesthetic sense comes into play in determining the arrangement of the elements in the initial strip and the use of the strip in creating the overall pattern. Combined with color and material variations, the possibilities for playing are vast. You will never need another rainy day project!

189

Above:
Fig. 5.16 A fifth generation unit from Fig. 5.14 forms a P4' group, with an additional unit overlaid in the center.

Below:
Fig. 5.17 The unit from Fig. 5.16 forms a P4 group. We will leave it to you to generate a plane group from this image!

Above and left:
Fig. 5.18   This image was originally hand created using a fractal-like approach to its construction. Copies of the pentagon are added at Golden Cuts of its sides. The unit was then tiled.

Below:
Fig. 5.19   This computer generated fractal uses a pentagonal initiator wherein each side is divided such that a portion in the center is replaced with the legs of a Golden Triangle. Note the Lutes of Pythagoras that appear at the vertices of the pentagon. It was developed using a component of the Geometry Grapher software created by Jonathan Choate for Houghton-Mifflin.

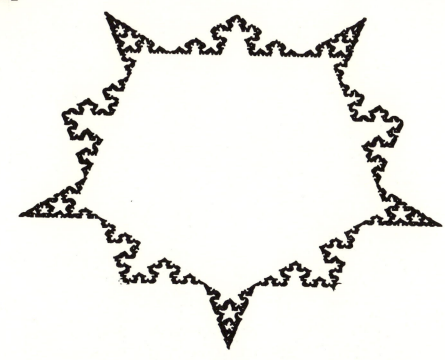

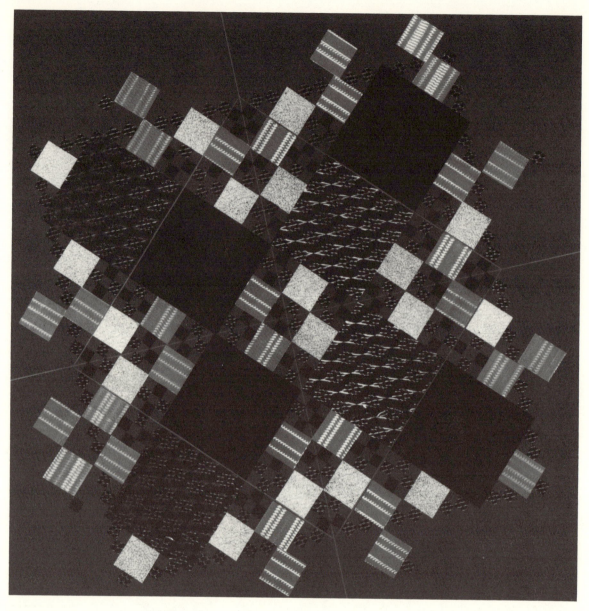

Generating Fractal Images on the Complex Plane

Some of the spectacular, colored fractal images you might have seen are created on the computer using a slightly different process than the one described in the preceding section. Rather than having a geometric figure act as the initiator, these images require an iterative function with a numerical initiator. Their complexity is limited by the number of units on the computer screen since they are colored pixel by pixel. Sections of the image can be viewed in minute detail because of the machine's ability to "zoom" into small portions of the plane. One function, alone, can offer an infinite variety of images depending upon which section of the plane is on the screen.

Remember that the computer screen is, in actuality, an invisible grid. Creating fractals requires locating points on this grid, and, therefore, we need a pair of axes. The Cartesian coordinate system allows us to plot *pairs* of real numbers on the plane. We are now going to use roughly the same game board, but we will alter the rules slightly so that we can plot *single* numbers on the plane. In spite of its familiar appearance, our new game board is called the **complex plane**. But, before we can make moves on the board, we must understand the playing pieces, the **complex numbers**.

The use of complex numbers arose when there was a need to solve equations such as $x^2 + 1 = 0$. This does not have a real number solution since to solve it we have $x^2 = -1$, and we know that the square of a real number, positive or negative, is a positive number. The solution is $x = \pm\sqrt{-1}$, and since it is not real we call it imaginary (but we are going to use it anyway!) In fact, the square root of any negative number is imaginary, but can always be expressed as the product of a real number and $\sqrt{-1}$.

For example:
$$\sqrt{-4} = \sqrt{4}\sqrt{-1} = 2\sqrt{-1},$$
$$-\sqrt{-17} = -\sqrt{17}\sqrt{-1},$$
$$\text{and } \sqrt{\frac{9}{16}} = \sqrt{\frac{9}{16}}\sqrt{-1} = \frac{3}{4}\sqrt{-1}.$$

Since $\sqrt{-1}$ is so important to the game, it has its own name, i, and the numbers in the previous examples, then, are written $2i$, $\sqrt{17}i$, and $\frac{3}{4}i$.

Let us look at the powers of i.
$i = \sqrt{-1}$
$i^2 = (\sqrt{-1})^2 = -1$
$i^3 = i^2i = -1i = -i$
$i^4 = i^3i = (-i)i = -i^2 = -(-1) = 1$
$i^5 = i^4i = (1)i = i,$
and the cycle repeats.

The powers of *i* take on only four values $\pm i$ and ± 1, which simplifies the arithmetic of complex numbers. But, what *are* complex numbers? They are numbers of the form *a + bi* where *a* and *b* are real numbers and *i* is the now familiar $\sqrt{-1}$. Thus, every complex number has both a real and an imaginary part.

Some examples of complex numbers, a + bi, are:

$3 - 2i$

$\frac{1}{2} + \sqrt{6}i$

$0 - 5i = -5i$

$4 + 0i = 4.$

When a = 0, as in the third example, the number is said to be purely imaginary. When b = 0, as in the fourth example, the number is a real number. Therefore, we can see that the **set** of complex numbers is large enough to contain the entire set of real numbers. Every real number is complex but not every complex number is real. For those readers new to complex numbers, their arithmetic is described in Appendix B.

The complex plane has two axes, but now the horizontal becomes the *real* axis and the vertical, the *imaginary* axis. Fig. 5.20 illustrates the differences in plotting points on the two.

Let us now consider an iterative

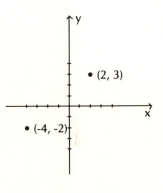

Real (Cartesian) Plane

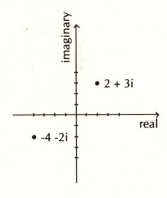

Complex Plane

Above:
Fig. 5.20 A comparison of the real and complex planes.

function on the complex plane: $f(z)=z^2+c$, where z and c represent complex numbers. If we begin by letting $z = 0$, and choose a value of c, say $2 + 3i$, we can then iterate the function by first squaring z and then adding the value of c, $2 + 3i$.

$f(0) = 0^2 + 2+3i = 2 + 3i$.
We now replace 0 with $f(0)$ to get
$F(2+3i) = (2+3i)^2 + 2 + 3i$
$= (2+3i)(2+3i) + 2 + 3i$
$= 4 + 12i + 9i^2 + 2 + 3i$
$= 4 + 12i - 9 + 2 + 3i = -3 + 15i$.
Iterating again,
$f(-3 + 15i) = (-3 + 15i)^2 + 2 + 3i$
$= (-3 + 15i)(-3 + 15i) + 2 + 3i$
$= 9 - 90i + 225i^2 + 2 + 3i$
$= 9 - 90i -225 + 2 + 3i$
$= -214 - 87i$.

Thank goodness the computer can do this type of thing faster and more happily! If we were to continue this iterative process, the values in each generation would travel farther and farther from the origin. The points obtained in each generation provide the framework for the **orbit** of the z,

Above:

Fig. 5.21 The first three numbers in the orbit of zero when $c = 2 + 3i$.

Below:

Fig. 5.22 The Mandelbrot Set.

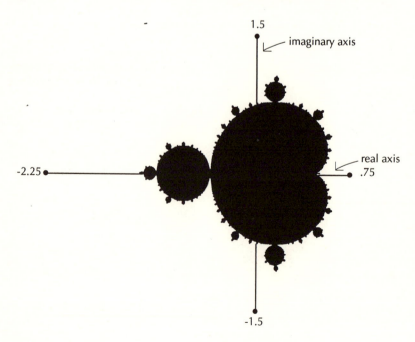

called the **seed**, in this case 0 (Fig. 5.21).

For most values of c the orbit of zero will shoot out to infinity. However, for some, the orbit will settle close to a particular number or will alternate back and forth between any number of values. We can separate the complex numbers according to what types of orbits zero has. This separation forms the basis for an exclusive club. A number becomes a member if the orbit of zero does *not* go to infinity. If we graph the members of the club (wearing formal black!), the result is as in Fig. 5.22. This select group is called the Mandelbrot set, discovered by Benoit Mandelbrot in 1980.

Fig. 5.22 shows the Mandelbrot set in its entirety. Notice the size. The real and the imaginary axes each travel to infinity in both the positive and negative directions. The set itself extends from about two units in the negative direction to less than one unit in the positive along the real axis, and takes up roughly two units on the imaginary axis. Its entire area in the complex plane could be likened to one speck of sand in the universe.

It may not be evident that the Mandelbrot set is an object of intense mathematical interest and intricacy. It is only when one "zooms" to a detail that its complexity begins to reveal itself. As in the natural world, it is at the boundary, along the edges, that unexpected behavior occurs. Any magnifications of the border region will display variation, and yet, imbedded in this variation is a kind of similarity. Even though it is not a true fractal, smaller and smaller versions of the Mandelbrot set appear along the intricate tendrils that spiral out on the edges of the set. This is a phenomenon

that can be observed again and again with each new magnification.

Associated with the Mandelbrot set is another group of sets called Julia sets, after French mathematician Gaston Julia. Remember that a complex number, c, belongs to the Mandelbrot set if the orbit of zero does not fly off to infinity. Instead, consider leaving c fixed and observe the orbits of all the different seeds, z, within certain parameters. If we choose only those z's whose orbits do not go to infinity, they form the Julia set of the number c. Julia sets are true fractals in that they exhibit self similarity. Fig 5.23 shows the Julia sets for a few values of c. Notice the visual differences between the sets for $c = -.194 + .6557i$ and $c = i$ in spite of the close proximity of these two values. Again we see evidence of sensitive dependence on initial conditions.

The Mandelbrot set has such special properties and has opened up so many areas of investigation that it is thought to be the most complex object in all of mathematics. However, there are functions other than $f(z) = z^2 + c$ that couple mathematical interest with visual excitement. Today, many other researchers are working in this field.

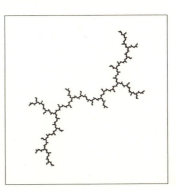

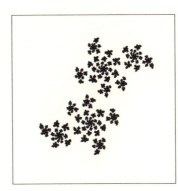

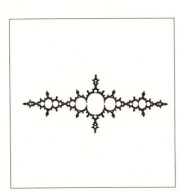

This page
Fig. 5.23 Julia sets for several values of c in the Mandelbrot set.
Clockwise from top:
c = -.194 + .6557i
c = i
c = .27334 + .00742i
c = .11031 - .67037i
c = -1.25i

Our feeling for beauty is inspired by the harmonious arrangement of order and disorder as it occurs in natural objects — in clouds, trees, mountain ranges, or snow crystals. The shapes of all these are dynamical processes gelled into physical forms, and particular combinations of order and disorder are typical for them.

Gert Eilenberger
German physicist

Part of the aesthetic appeal of these images is due to the addition of color. Most artists choose color for its emotional impact, while mathematicians might choose color for informational purposes. In fractal images, most often, black is used to color those pixels that represent complex numbers that belong to the set. Based on a previously determined scheme, other points are variously colored to indicate how *quickly* the orbit of zero moves to infinity. Experimental mathematicians now feel that a sophisticated art form has been born, and with that realization comes the desire for a child that is both beautiful and rational.

Even though the idea of coloring pixel by pixel is a relatively new one to the mathematical community, artists have been exploring this technique by hand for centuries. Ancient mosaics give testimony to the fact that the idea of working unit by unit is not a new one. Woven tapestries grow by weft yarns covering single warp yarns, one at a time. The Jacquard loom required cards punched unit by unit, and it was the paintings of the French artists of the last century that applied this idea to techniques on canvas. The painter Georges Seurat made this approach the essence of his art. While the daily activities of his countrymen were the subject matter for his works, and his compositions were derived from the use of the Golden Ratio, his technique of point by point color was embedded in his desire to capture with inert pigment the brilliance of natural light.

The scientific community is excited about the potential of fractal geometry. It has not, however, given up the geometry of Euclid. Rather, it looks at both as useful tools for modelling different aspects of reality. Regular objects are still best explained through classical geometry, while irregular natural forms are better scrutinized through this contemporary mode.

Fractals look to explain chaotic conditions. But the underlying assumption is that even chaos has its rules, and the complex is accessible to systematic study. It is a better way to look at the many "bits" of nature in order to understand the relation of the parts to the whole. Fractal geometry is a finer sieve in which to sift reality. Yet, it is still only a sieve. It is not Reality itself. Let us not fall so much in love with the simulations that we forget their origins. If fractals are changing our understanding of the world, let us fervently hope that the change will be for the better.

Below:
Georges Seurat, French, 1859-1891. Sunday Afternoon on the Island of La Grande Jatte. Oil on canvas, 1884-86. 207.6 x 308 cm. The Helen Birch Bartlett Memorial Collection. 1926.224. Photograph copyright 1991, The Art Institute of Chicago. All rights reserved.

Although the color areas appear solid, the entire surface of this work is covered by a multitude of dots of color joined into unity by the human eye.

To analyze this painting, a starting place might be to notice that the butterfly falls on the intersection of a vertical and a horizontal line, each connecting Golden cuts of the sides of the framing rectangle. Try to carry the analysis further, knowing what you know about harmonic armatures (See Chapter 4, Book 1, Universal Patterns).

Problems

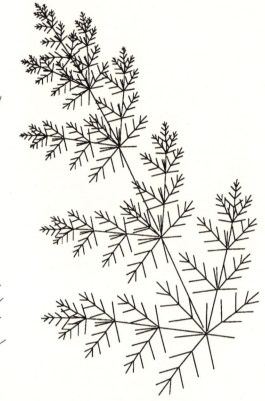

Above:
Fern fractal, suggestive of a natural form, was developed on Fractasketch, *a program by Dynamic Software.*

5 Construct any polygon. Trace it onto tracing paper. Remove the original and trace the tracing. Take this process through to the sixth generation. Compare the final result to the original. In writing, respond to the following questions. Are there differences? If so, what are they and how do you explain them? What would this suggest about the natural world and/or human made decisions, structures, etc?

6 Plot and name points on the *complex plane* that would correspond to the following Cartesian points: (4,1), (3,0), (2,-5), (0,-6).

7 Develop a generator based on an initiator that is a line segment. Develop another by altering it using a symmetry operation. Take each through four generations to observe the differences. If using a computer, carry out to many more generations.

8 Create a fractal subdivision of an equilateral triangle, a square, or a regular pentagon. Carry it out to the fourth generation if working by hand, more if using a computer.

1 Create a 4th generation fractal that begins with a line segment that branches at both ends such that r = 1/2 and $\theta = 60°$.

2 Let $f(x) = x^2 - 3$. Begin with x = 0 and iterate 8 times. (Aren't you glad you own a calculator!)

3 Let $f(x) = x + 4 - 2i$. Begin with x = 0 and iterate 10 times.

4 Make a fractal subdivision of any special triangle using midpoints as in Fig.5.13. Iterate four times if working by hand, but more if using a computer.

Projects

1 Research the topic of tree design. Why do different species have different shapes, heights, type of leaves, flowers, etc? Support your writing with drawings, photocopies, photographs, etc.

2 Use the results of Problem 1 as the inspiration for an artwork in two dimensions.

3 Illusion Cube

In the interior of a square, use one of the algorithms described in the chapter, or define your own, duplicate it six times, form a cube, with flaps for gluing. Add personal touches of color, texture, pattern, surprises, etc. before putting the pieces together.

4 Using Problem 4 as an armature, create an artwork. Incorporate natural elements.

5 Use the ideas of Problem 5 to create an artwork that takes an image through a sequence of distortions.

6 Using what you have learned through Decat's adventures, develop a walk that is described through the use of colored threads and nails on a piece of plywood. Let each thread color be a different generation. Be concerned with the aesthetic qualities of your solution.

7 Create a fractal pillow, quilt unit or piece of clothing using the fabrics of your choice.

8 Create a monohedral or dihedral tiling with fractal polygons. Add three colors in either an ordered or a random fashion. Do not be afraid to alter the boundaries.

9 Choose one of the following:

a. Develop a fractal pattern on an equilateral triangle. Make 20 multiples and construct an icosahedron.

b. Use the same process for creating twelve regular pentagons and construct a dodecahedron.

c. Use the 20 triangular units plus the 12 pentagonal units to construct an icosadodecahedron. The vertex net is shown in Fig. 5.24. Every vertex should be the same in a uniform polyhedron.

Make sure you consider the aesthetics of any of the above.

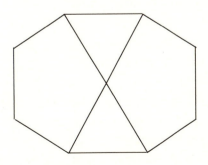

Left:
Fig. 5.24   Vertex net for the icosadodecahedron.

10 Choose a branching fractal structure. Use it as a basis for an artwork that deals with the concept of the four seasons; moving from the "naked" winter branches through spring, summer, and fall. Use materials of your choice.

11 Use branching patterns that you have constructed in order to create a variety of flora. Develop an imaginary landscape in materials of your choice. The photocopier can enlarge or reduce, or multiply your initial units. Consider using colored tissue papers, wallpaper, contact paper, tracing paper, colored markers and pencils, etc.

12 Bake a batch of Mandelbrot (Mandelbrot means almond bread) and share it with your friends without getting fractal crumbs all over the carpet!

Fibonacci Meets Mandelbrot

3 eggs
1 cup sugar
1 cup oil
1/2 tsp. lemon extract
1 tsp. vanilla
2 tsps. baking powder
3.5 cups flour
1 cup white raisins
1 cup slivered almonds
Preheat oven to 350°.

In a large mixing bowl beat sugar, oil, eggs, vanilla and lemon extract. Mix together until smooth.

Mix flour and baking powder together. Add to above mixture

Stir in raisins and nuts.

Shape into two logs and place on a non-stick baking sheet.

Bake for 30 minutes or until lightly browned.

Slice logs. Place pieces back onto the baking sheet. Bake again until lightly browned on both sides.

Enjoy!

Further Reading

Barnsley, Michael. *Fractals Everywhere*. Boston: Academic Press, 1988.

Gleick, James. *Chaos: Making a New Science*. New York: Penguin Books, 1987.

Mandelbrot, Benoit. *The Fractal Geometry of Nature*. New York: W. H. Freeman and Co., 1983.

McGuire, M. *An Eye for Fractals: A Graphic/Photographic Essay*. Reading, MA: Addison Wesley Publishing Co., Inc.,1990.

Peitgen, Heinz Otto and P. H. Richter. *The Beauty of Fractals*. New York: Springer-Verlag, 1986.

Peitgen, Heinz Otto and Dietmar Saupe. *The Science of Fractal Images*. New York: Springer-Verlag, 1988.

Peterson, Ivars. *Islands of Truth*. New York: W. H. Freeman and Co., 1990.

Peterson, Ivars. *The Mathematical Tourist*. New York: W. H. Freeman and Co., 1988.

Stewart, Ian. *Does God Play Dice? The Mathematics of Chaos*. Cambridge: Basil Blackwell, Ltd., 1989.

6 *The Pliable Plane*

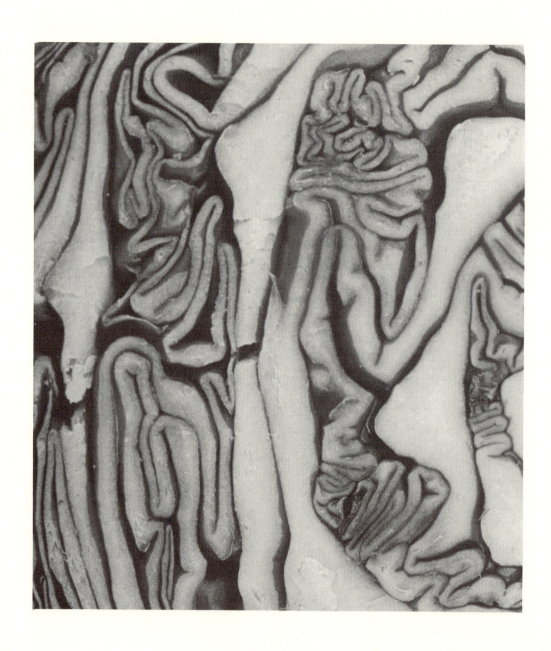

Below:
Richard Newman. System Re-
peat. 48" x 96", √4 Rectangle.
Acrylic paint on plywood. This
painting is one of a series of
works using the same 18 grad-
uated circles in a line. The algo-
rithm for generating this image:
five lines in one direction, re-
verse, five lines in the other.

An artwork is a function of infinitely many variables. Intuition is the juggling act of being able to deal with all the variables simultaneously. It is the sensitivity of the juggler that accounts for the quality of the work.

Optical Bending

To this point we have considered the plane to be both flat and rigid. If we make the assumption, however, that it is pliable and capable of being stretched and bent, new possibilities open up. We begin to move from two dimensions into three, but still keep the plane essentially frontal. In this chapter our investigation of dimensionality will be minimal, leaving a more thorough exploration of the third dimension for Book 3.

Remember that, although not defined mathematically, we assume the Euclidean plane to have the attributes of infinite length and width, but no thickness, and that it is flat. Therefore, when we speak of a bounded portion of the plane which is buckled or bent, we are dealing with what mathematicians call a surface. For the artist, any surface can have illusions of depth so that it functions as an implied space as well.

In the natural world, all forms are three dimensional. They have length, width and thickness. Many forms, such as leaves, butterflies and tree bark, appear to be flat. When looked at closely, however, their surfaces abound with bulges, bends, crinkles and folds. In short, they possess all of the elements of the fractal quality of reality. Some of the forces that may change the two dimensional surface occur when: there is stress on the form; there is a need for more surface area within a given space; there is a need for the reinforcement of a form; or there is a need to increase the strength of a form while reducing its mass.

This page:
Pliability exhibited in different types of leaves.

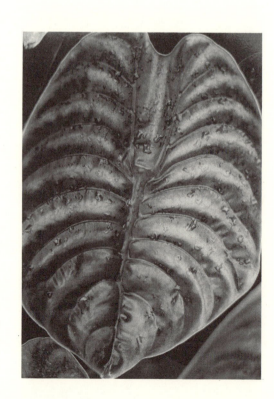

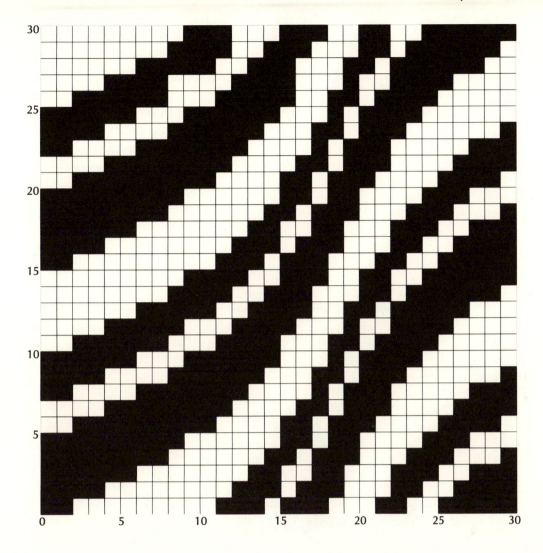

Fig. 6.1 Illusions of rippling on a portion of the Cartesian plane.

Here we will work with a restricted area of the plane, and define its surface through the use of a grid. It will act as an armature to be manipulated, one that is devoid of natural content. We shall begin with the basic square grid. For ease of handling, let us assume that our grid is oriented in such a way as to fall within the first quadrant of the Cartesian plane, where both *x* and *y* coordinates are positive. Let the boundaries be defined by the points (0,0), (0,30), (30,0), and (30,30). See Fig. 6.1. One way to alter the appearance of the plane surface is to fill in units on the grid in some systematic fashion so that the surface appears to ripple. This type of "bending" is actually the result of creating illusions of three dimensions on a two dimensional surface. It relies on the viewer's acceptance of information about spatial cues originally received through his or her real interactions with the natural world.

It is assumed that the surface plane, which the artist calls the picture plane, is parallel to the viewer and that it is seen straight on. Any mark made on it will activate the implied space of the plane, forcing the viewer to make decisions about elements within that space. These elements deal with the concepts of near and far, small and large, up and down, in front of or behind. Altering the surface affects perception.

In Fig. 6.1, the grid unit remained constant, and a single variable (shading) was introduced. With the addition of more variables, the surface becomes more activated. If we add a second variable, such as changing the shape of the grid units, a greater feeling of movement is achieved. The grid can be uniformly or non-uniformly stretched along either or both axes. In addition, it can be sheared, which alters the right angle relationship. Fig. 6.2 shows how our first example is transformed when played out on a variety of grids, still bounded by the points (0,0), (0,30), (30,30) and (30,0).

Below:
Fig. 6.2   Rippling effect on manipulated grids. In this figure there are five 30x30 grids adjacent to one another.

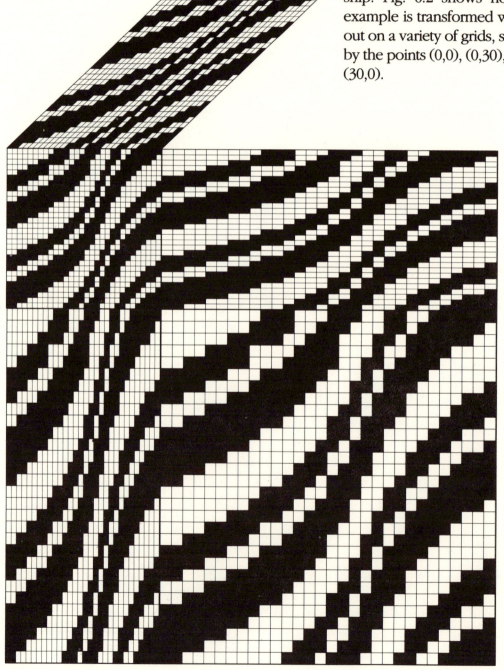

The whole field of fiber art, which includes baskets, tapestries and textiles, is predicated on the notion of physically interlacing elements that are suggestive of interactions on the grid. Elements that are fragile, in and of themselves, gain strength from the physical interlacing process such that the entire web is greater than the sum of its parts.

The basis for all interlacing is the concept of "under-and-over". Variety is obtained through the number of units chosen to lie either under or over other units. The most fundamental structure, the simple "under one-over one" model, also provides the greatest strength. However, humans, rarely content with mere repetition, seek to stretch limits, and in the process discover new structures. This has certainly been the case in the history of fiber, attested to by the myriad of patterns developed over the centuries.

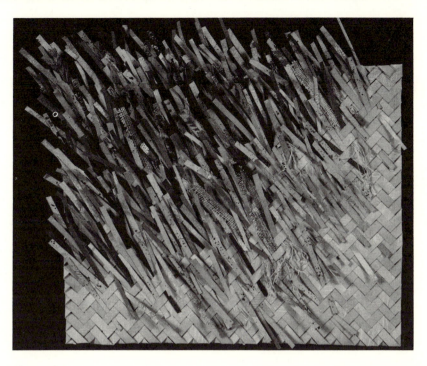

Above:
Sandy Weisman, Winter Morning. 15" x 19", paper, acrylics and linen, 1987.

Far left:
Linda Yackly, traditional basket form.

Near left:
Kay Rosenberg. Collapsible, Expandible Grid. 54" x 72" x 54", closed cell neoprene sponge, wood.

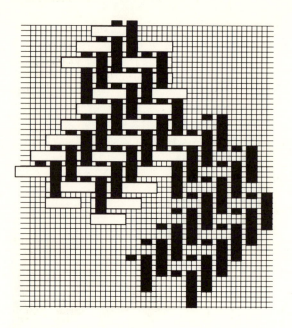

Fig. 6.3 Different interlacing patterns can be expressed using rectangular and isometric grids.

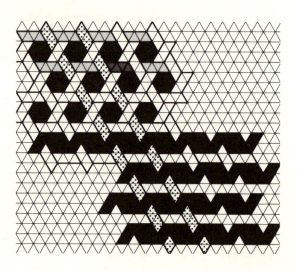

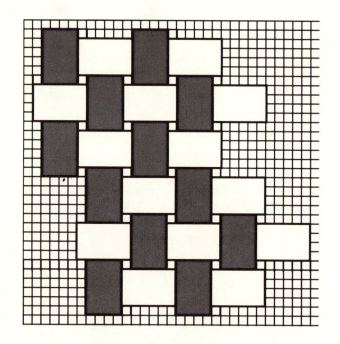

In reality, interlacing is a three dimensional activity. However, its illusion can be expressed on a flat surface through the use of a grid. The resulting pattern is dependent upon both the choice of the grid and the way it is manipulated. The sense of three dimensionality can be intensified with appropriate shading, as in Fig. 6.3.

Other illusions of pliability can be created once the grid has been determined. The interior of each of its units can be further altered in an infinite number of ways. The most basic move would be to introduce the diagonal which affords the illusion of curving. Fig 6.4 illustrates the changes that can occur using one or more diagonals and shading.

This page:
Fig. 6.4   The source for this group is a grid that shows fractional divisions of a line segment. Possibilities abound when the structure is combined with shading and symmetry operations.

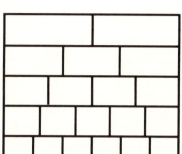

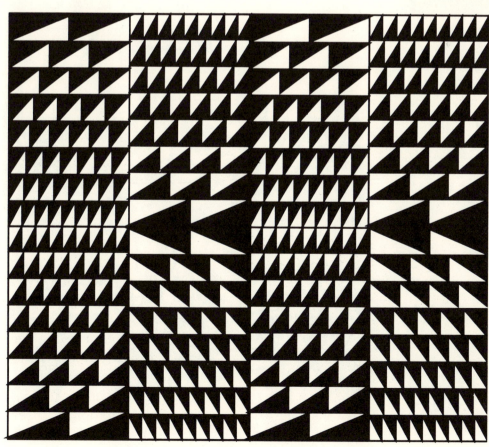

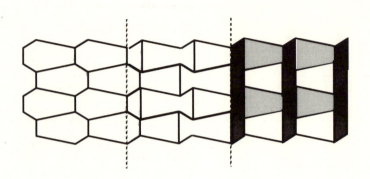

Ideas from previous chapters provide a wealth of information, techniques and guidelines for developing structures that, when coupled with shading, produce images that can bend, fold, warp or twist. Fig 6.5 shows how simple hexagonal tiling structures can be made to appear pliable.

Above:
Fig. 6.5   Pliability on hexagonal tilings.

Below:
Tiling using a student created unit which was based on Fibonacci divisions of a square. Brian Campbell.

Pliability on the Triangular Grid

In and of itself, the isometric grid does not appear pliable. However, when all three axes are used to develop the image of a rectangular prism, the feeling of three dimensions is very strong even without shading. This is due to the fact that figures containing angles other than right angles appear to have sections that advance or recede. See Fig. 6.6.

The most basic rectangular prism is a cube in which lengths along the three axes are equal. The next "construction" shows how you can easily develop the image of a cube on the triangular (isometric) grid.

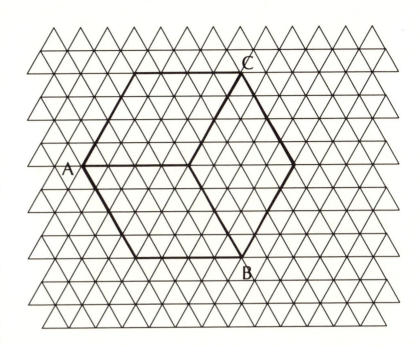

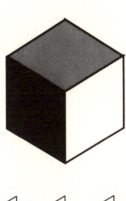

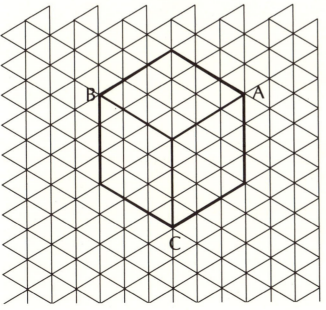

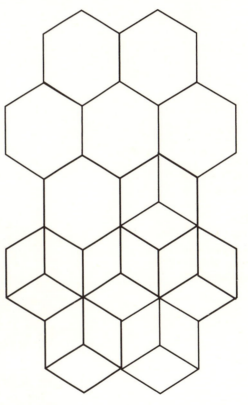

This page:
Fig. 6.6 The development of a cube on the isometric grid.

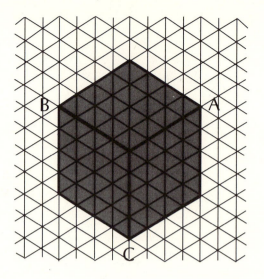

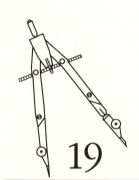

19

Draw the Image of a Cube on an Isometric Grid

Instant cube —just add line segments!

Given an isometric grid.

1. Choose the number of units that determines the length of an edge of the cube (in our example, 4).

2. On the grid, draw a regular hexagon with sides of the length determined in Step 1.

3. Label three nonconsecutive vertices A, B and C. These can be in either orientation as illustrated.

4. Draw segments from A, B and C that intersect at the center of the hexagon.

Now the figure appears to be a cube.

When multiples of cubic images are joined and stacked, units can be developed which have the appearance of dimensionality. Fig. 6.7 illustrates how using the isometric array, as opposed to the grid, simplifies the procedure by avoiding visual clutter. Fig. 6.8 shows how the single unit can be tiled so that the resulting pattern denies the flatness of the plane.

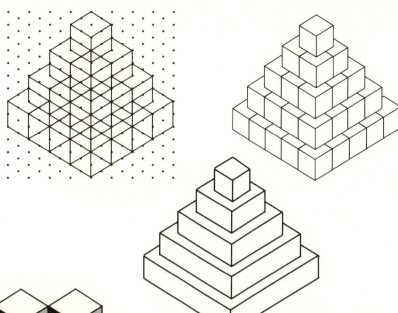

Above:
Fig. 6.7   Using the isometric dot array to create the illusion of stacking cubes.

Left and below:
Fig. 6.8 Tiling with cubic units.

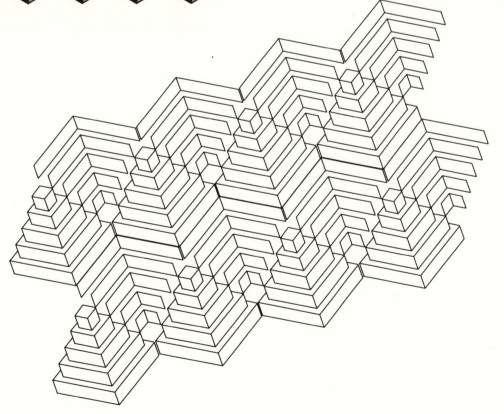

213

If the cube is not the rectangular prism of choice, an infinite variety of others are available. This construction shows how to create them with simple extensions of both thought and line.

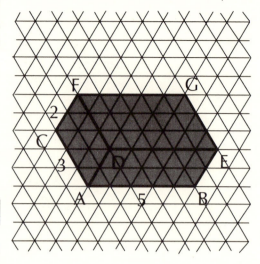

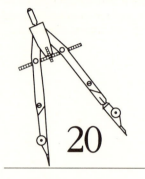

20

Draw the Image of a Nonregular Prism on an Isometric Grid

Given the isometric grid.

1. Determine the length, width and height* of the rectangular solid you wish to construct. Our example uses Fibonacci numbers 5, 3 and 2.

2. Choose a point (A) to serve as one vertex and draw three segments from it in different directions as shown to act as the edges of the solid.

$\overline{AB}$ is 5 units long (length).

$\overline{AC}$ is 3 units long (width).

$\overline{AD}$ is 2 units long (height).

3. Draw $\overline{DE}$ 5 units long and parallel to $\overline{AB}$.

4. Draw $\overline{EB}$.

5. Draw $\overline{DF}$ 3 units long and parallel to $\overline{AC}$.

6. Draw $\overline{FC}$.

7. From F and E draw segments parallel to $\overline{AB}$ and $\overline{AC}$, respectively, that intersect at G.

8. Shading heightens the illusion of three dimensionality.

Now the figure has the appearance of a rectangular prism.

* Length, width and height are purely arbitrary designations to describe different edge lengths on a rectangular prism. Either of the two prisms below could also be thought to have length, width and height of 5, 3 and 2, respectively. Position in the plane is arbitrary as well.

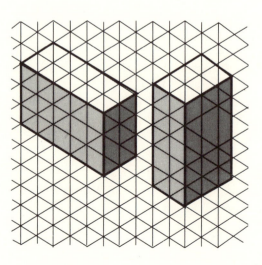

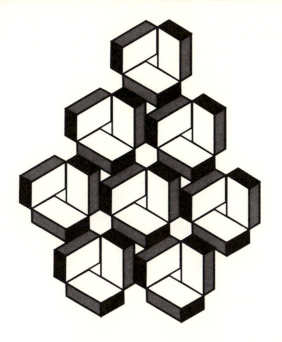

This page:
Fig. 6.9   Variations on units drawn on the isometric array. The illusion of dimensionality is enhanced by using multiples or combinations of the units.

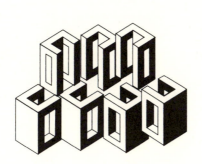

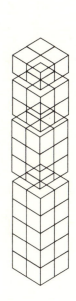

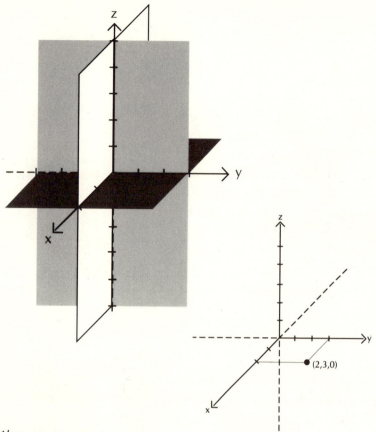

Above:

Fig. 6.10 *The coordinate axes in 3-space.*

Right and below:

Fig. 6.11 *To graph a point in three dimensions, say (2,3,5), begin by locating 2 on the* x *axis and 3 on the* y *axis. Through each of these points draw lines parallel to the other axis, and extend them until they intersect.*

Locate 5 on the z *axis. Draw segments parallel to the other axes as before.*

Through each of the points of intersection, draw a line that is perpendicular to the plane containing the point. The intersection of these three lines is the point (2,3,5).

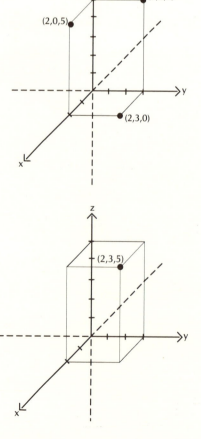

Plotting points in three dimensions uses the concept of the rectangular prism as its foundation. Mathematically, graphing points in 3-space is a logical extension of the same activity in 2-space. Rather than using ordered pairs to locate points on the plane relative to two axes, ordered triples, (x,y,z), are used. They designate points in space as they relate to three axes, each of which is perpendicular to the other two.

To work with this idea on the plane of the paper, the y and z axes lie in the plane and the x-axis is drawn as if it pierces the plane at right angles. The point of intersection is still called the origin and has coordinates (0,0,0). Each axis is calibrated so that the positive real numbers lie on the portions drawn with solid lines and the negative real numbers lie on the parts indicated with dotted lines (see Fig. 6.10). The system of axes determines three planes that divide space into eight separate parts known as **octants**. Fig. 6.11 shows how to graph a point in 3-space.

In the field of analytic geometry, it is possible to travel beyond even the limits of our real three dimensional space. As with the preceding example, it is not necessary to reinvent the wheel for each new dimension. Rather, the rules are developed using logical extensions of concepts in the previous space by introducing a new variable. Higher dimensions actually represent intellectual spaces rather than physical ones. However, there are many practical applications of higher dimensional geometry in our three dimensional world.

Ambiguity, Contradiction and Pliability

We have just looked at some images in which the unit appears to be stable. However, with certain kinds of alterations, figures with optical ambiguity can be constructed. These put the viewer in an uncomfortable situation in which objects seem to shift in orientation and become unstable. For example, in Fig. 6.12 the position of the smaller prism can be interpreted in several different ways, none of which can be considered correct or incorrect.

Interest in this phenomenon in the Twentieth Century may be linked to ambiguity in other areas of our lives. This representation becomes a visual metaphor for the uncertainty of our times. Fig. 6.12 and Fig. 6.13 demonstrate how ambiguous figures can be tiled to create visual pliability on the plane (and disturb you emotionally at the same time!).

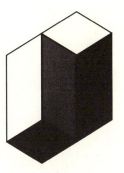

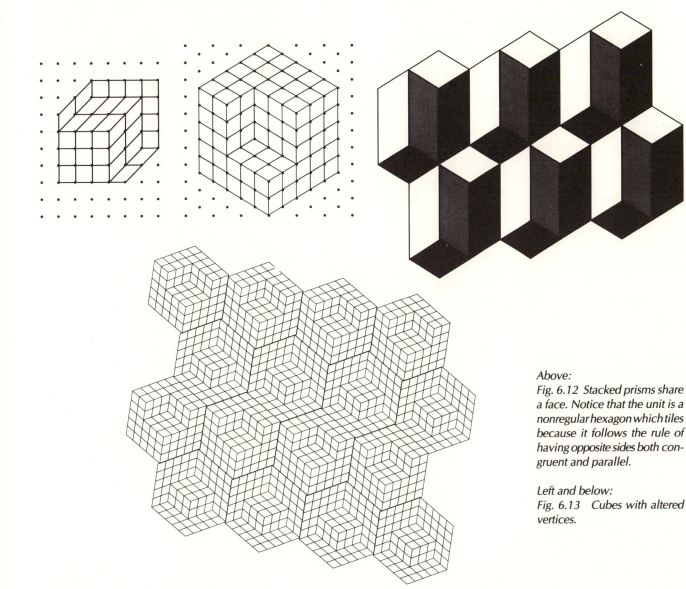

Above:
Fig. 6.12 Stacked prisms share a face. Notice that the unit is a nonregular hexagon which tiles because it follows the rule of having opposite sides both congruent and parallel.

Left and below:
Fig. 6.13 Cubes with altered vertices.

By altering portions of possible figures, contradictory ones can be designed. If they are developed within polygons that can be tiled, the resulting surface may have both pliability and a feeling of tension caused by the contradiction. In this case, as before, the eye attempts to resolve the ambiguity, but to no avail. All interpretations are denied by the inherent impossibility of the figure.

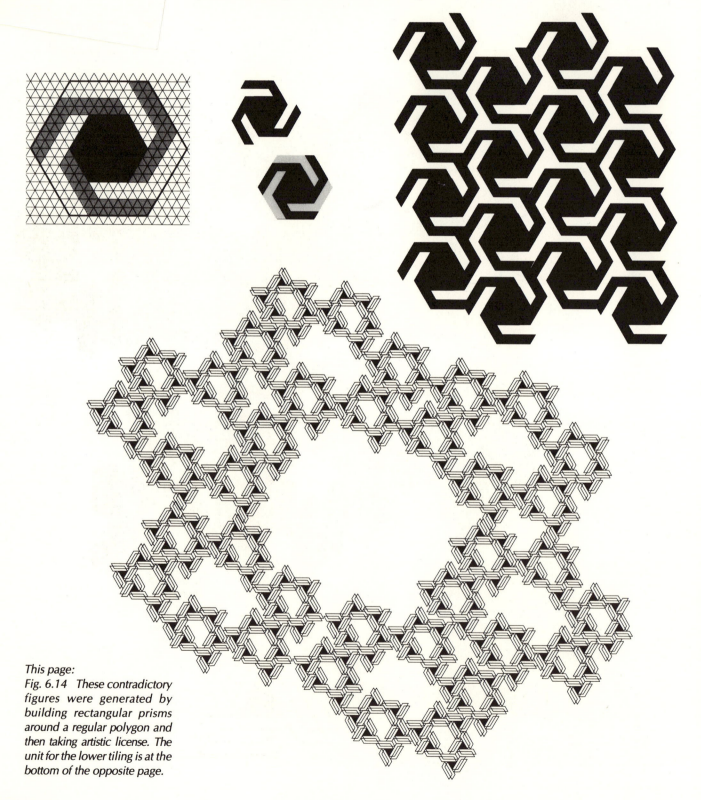

This page:
Fig. 6.14 These contradictory figures were generated by building rectangular prisms around a regular polygon and then taking artistic license. The unit for the lower tiling is at the bottom of the opposite page.

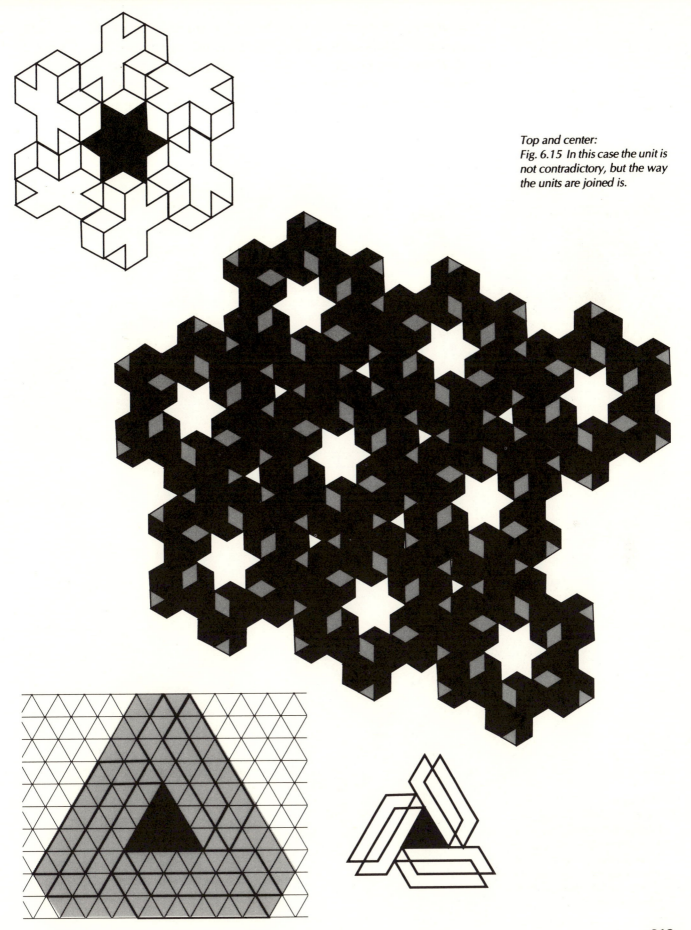

Top and center:
Fig. 6.15 In this case the unit is
not contradictory, but the way
the units are joined is.

Fig. 6.16 *The illusion of pliability results from replacement of straight segments with curves.*

Anamorphosis: Distortions for Analysis and Synthesis

In the previous section, we have seen how manipulating figures on a constant grid can produce illusions of dimensionality. Here we will explore distorting the grid itself to obtain similar illusions. This process takes a figure away from the hard edged, straight line abstraction to one that is more organic in character. The effect is achieved by replacing certain line segments with curves. Fig. 6.16 illustrates the process on a square grid. The construction following describes a technique for creating a bubble-like bump in the plane when more regularity is desired.

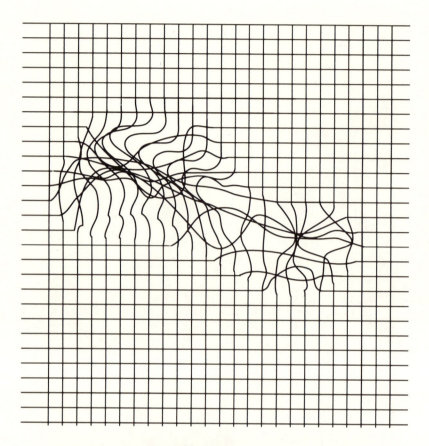

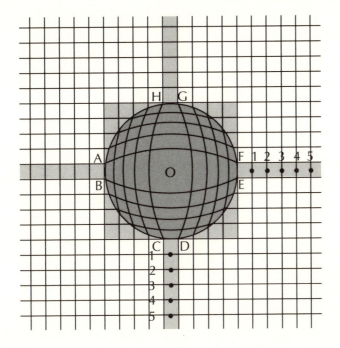

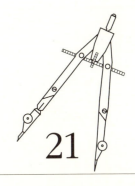

21

Construct a Bubble in the Grid

Given a square grid

1. On the grid, draw a square whose sides have an odd number of units, in this case nine. The sides will serve as the boundaries of the bubble.

2. Inscribe Circle O in the square using the method described in Fig. 3.3, page 91.

3. Lightly shade, outside the circle, the columns and rows that intersect at O. Label the four pairs of points where these intersect the circle A, B, C, D, E, F, G and H, consecutively.

4. Eliminate the grid lines from the interior of the circle.

5. The arcs within the bubble will be constructed from the center out, using the lines on the grid as endpoints, and reducing the length of the radius each time, until it approaches that of the original circle. The following steps give the specific arcs in our example*. All line segments mentioned will be those

in the rows and columns shaded in Step 4.

6. Place the metal tip in the exterior of the circle on the midpoint of the line segment 5 units away from $\overline{CD}$ and draw arc AF.

7. Without changing the setting, and staying 5 units away from the square, move around in a counterclockwise direction, and draw arcs CH, EB, and GD.

8. Place the metal tip on the midpoint of the segment 3 units away from $\overline{CD}$ and draw an arc whose endpoints lie on the next line out from the center after $\overleftrightarrow{AF}$. Continue around the circle as before.

9. The remaining two sets of arcs will be centered on the midpoints of the segments 2 units and 1 unit from $\overline{CD}$, respectively, and drawn as before.

Now the figure is a bubble in the grid.

* Variations in the arcs could be gotten by choosing centers at different distances from the center of the circle.

A similar technique can be used on an object that already has the illusion of three dimensionality, such as the cube in this construction.

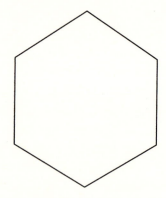

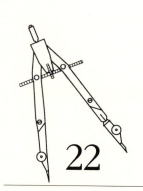

22

Draw a Cut-away Cube

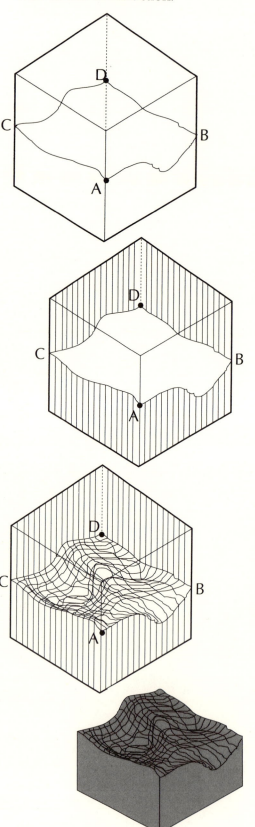

Given a regular hexagon*

1. Lightly indicate the segments that give the hexagon the appearance of a cube.

2. Choose an arbitrary point (A) on the edge that appears closest to you. This will determine where the cut begins. Draw a curve the shape of the portion to be cut away from A to C and from A to B.

3. Choose an arbitrary point (D) on the edge that appears furthest from you. Draw curves from D to C and D to B.

4. Draw vertical grid lines on both front and back faces.

5. Connect endpoints of verticals on "front faces" to corresponding endpoints of verticals on opposite "rear faces" using curves in the shape of your choice.

Now the figure is a cut-away cube.

This technique would also work on any of the drawings of rectangular prisms that we have examined.

To this point our discussion of pliability has focused on purely abstract elements. Now let us examine ways in which these abstractions can be applied as an investigative tool for analyzing shapes in the natural world. Morphology is the study of the form of material things and the physical forces that alter them. A morphologist is particularly interested in the connection between form and function of natural structures, the evolution through time of particular species, and the similarities in forms of members of families.

Using the grid, with its potential for myriad distortions, provides the morphologist with a structure for analysis. The activity of transforming the Cartesian coordinate plane falls within the domain of **anamorphism**: the transformation of images or natural forms in evolution. The alternative grids produced by the manipulation of the Cartesian plane allow a morphologist to plot position and order so as to be able to compare seemingly dissimilar forms.

The process can be illustrated through the use of a plane image of the human face. The first quadrant of the Cartesian grid is overlaid and the coordinates of key points are noted. A different grid is introduced and the same points are plotted on it. Each unit is then transferred, one at a time, to complete the distorted image. If there is difficulty in transferring all the necessary information, the units can again be subdivided, much like the zooming process on the computer screen.

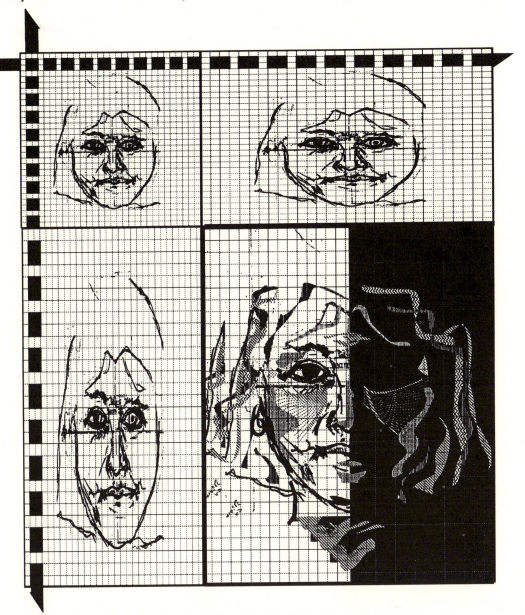

Below:

Fig. 6.17 Self-portrait transformation. Cynthia Hastings. A scanned hand done drawing was integrated with a grid and then manipulated on the computer.

Above:
Fig. 6.18 Butterfly transformation.

On these pages, this idea is played out using other natural forms. Once again, the interest is in relationships among apparently dissimilar structures. For example, species of fish look very different from one another, and yet each has common features such as eyes, mouths, fins and tails, usually located in roughly the same orientation to one another. It is the change in distance between features that alters the overall proportions, thus making one species seem unlike another. Unity and diversity exist in all things.

Below:
Fig. 6.19   Fish transformation.

The biologists say that all morphology is adaptive, meaning that through generations, a species will alter its form to better suit its climate, terrain, movements, food intake, fighting, mating, and all the countless circumstances that constitute its environment and its living within that environment—its functioning. The artists, designers and architects have put it another way—form follows function, meaning that the form of an object should be obedient to the necessities of its function.

Both statements mean about the same thing applied to the natural and the human environments, but both lead us to believe that there may be an end result, when in fact the process itself is the end, and the object of this process, the evolving form, could be forever changing, hoping to catch up to that elusive form of perfect obedience.

Christopher Williams
Origins of Form

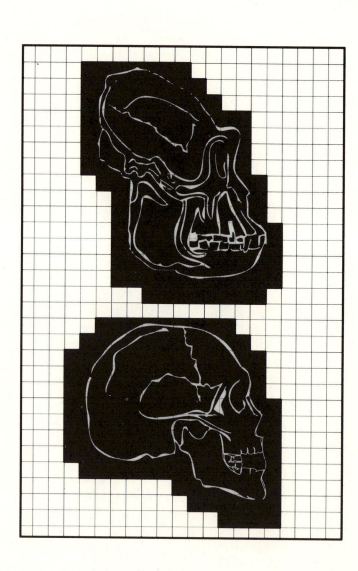

Above:
Fig. 6.20 Transformation from
Ape skull to Homo sapiens skull.

Problems

1 Using the isometric grid, develop a rectangular prism having length 3, width 5 and height 8. Repeat two times, interchanging the dimensions of the length, width, and height.

2 Use a square grid, 50 units by 50 units, and create an image that appears pliable. Use only black and white.

3 Do a rubbing or tracing, or find a photograph of a symmetric leaf. Superimpose a polar grid onto it. Deform half the image by stretching the grid and join these two halves into a new leaf form.

4 Create a motif within a square. Stretch the original through one of the Dynamic or Ø-Family Rectangles. Create a dihedral tiling with the two units.

5 Create an ambiguous or contradictory figure that was not illustrated in the text.

6 Plot the points (1,2,5), (2,4,-1) and (-2,6,-4) using the techniques described in Fig. 6.11.

7 On the same grid, construct both a bubble and an irregular curved section.

8 Draw a cut-away rectangular prism (not a cube) that has at least three peaks in the curved surface.

9 Using a photograph of a human face, overlay a square grid and note key points. Transfer the image to a grid composed of Golden Parallelograms.

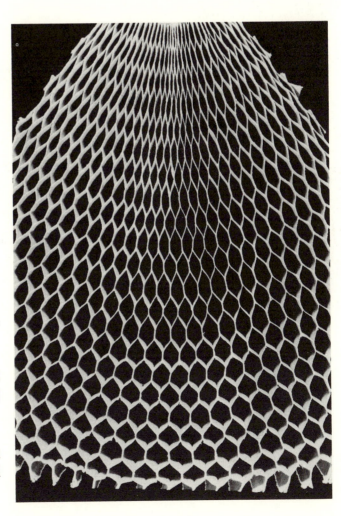

Below:
Flexible paper grid illustrates pliability.

Projects

1 Use any of the problems from this chapter as the basis for a project. Be concerned with quality, craft and creativity.

2 Using the isometric grid and what you have learned about drawing prisms, construct an imaginary city-in-space landscape. Use color and techniques that would suggest an otherworldly environment.

3 Cut any heavyweight paper into strips of differing Fibonacci widths. Interlace using the concept of "under-and-over". Pull up various portions of the strips to obtain an actual change in surface. Mount for display as a wall scupture.

4 Find three different photographs either in black and white, color or a combination. Construct an harmonic grid on top of each photograph either directly or by laying over a piece of tracing paper. On good quality drawing paper or colored construction paper or canvas, construct an harmonic grid that is twice the size of that on the photographs. Now, using the medium of your choice (such as colored pencils, markers, acrylic paint) transform units from the first photo onto the upper third of your newly constructed grid surface, switch to the second photo for the second third of the work, and use the last photo for the bottom third. Look to see what adjustments have to be made to the final image to create artistic unity. Then do it.

5 Find a black and white photo. On good quality materials, construct a Dynamic Rectangle grid. Transfer units from the original photo, but leave every other unit blank or use a Fibonacci progression in relating blank to filled areas. Fill in the blank units with color(s) and materials of your choice.

6 Develop a study for a seven panel folding screen by using a $\sqrt{7}$ Rectangle and its reciprocal (See Chapter 4, Book 1, *Universal Patterns*). The theme of this screen will be Nature in which the elements are to be butterflies, leaves, birds, and grasses. Use materials of your choice.

7 Using any grid and multiples of an alphabet letter, or your own initials, developed within the interior units of your grid, create an interesting black and white design.

8 *Self Portrait*

a. Tape a piece of gridded acetate to a mirror. Stand at a fixed distance from the mirror so that your face fits comfortably within the grid. Note the coordinates of key points of your face. Subdivide a piece of quality paper using the same grid as that on the acetate. Transfer the points to it. Use the results to create a self portrait.

b. Transform the grid in some manner. Use the same points to create a "self-portrait mask" and finish it in any way you choose.

Below:
Pliability in concrete and aggregate wall relief.

9 Begin with an image of an actual parrot or other colorful bird, and transform its shape in at least three ways. Focusing on the results of the last transformation, execute a fantasy bird in drawing, collage, fabric, assemblage, etc.

10 Choose a photograph of an animal face. Photocopy it a minimum of three times. Slice each copy into horizontal, vertical or diagonal strips. Arrange the strips leaving space between to fill with drawing. Mount together on a single board.

11 Using the ideas suggested by the Noodle Illusion Box in the photograph, create your own illusion structure in materials of your choice.

Richard Newman. Noodle Illusion Box. *Photo sculpture.*

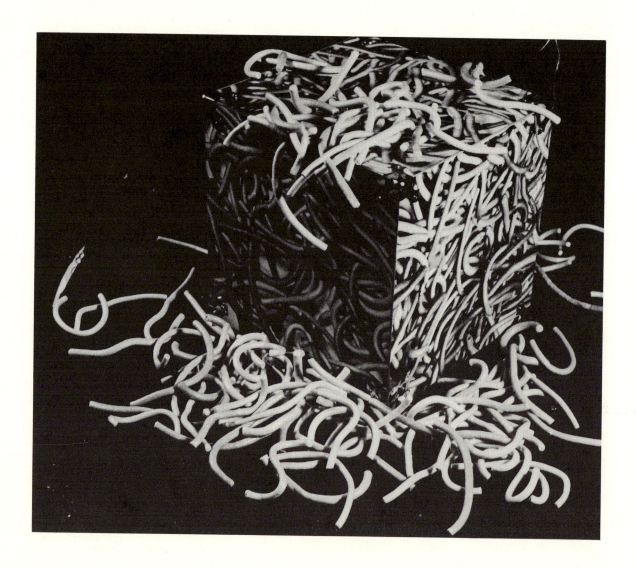

Further Reading

Locke, John. *Isometric Perspective Designs*. New York: Dover Publications, Inc., 1981.

Lord, E. A. and C. B. Wilson. *The Mathematical Description of Shape and Form*. New York: John Wiley and Sons, 1984.

Rowell, Margit. *The Planar Dimension*. New York: The Solomon R. Guggenheim Foundation, 1979.

Thompson, D'Arcy. *On Growth and Form*. Cambridge: The University Press, 1969.

Turner, Harry. *Triad Optical Illusions*. New York: Dover Publications, Inc., 1978.

Wade, David. *Geometric Patterns and Borders*. New York: Van Nostrand Reinhold Company, Inc., 1982.

Williams, Christopher. *Origins of Form*. New York: Architectural Book Publishing Company, 1981.

Conclusion

A really good question leads to other questions. It opens up areas of investigation rather than narrowing them. A really good question might not be answered easily even if, on first asking, it seems simple. It may take centuries.

We encourage you to be askers of questions . . . and seekers of Truth!

And watch where you step!

Selected References

Arguelles, Jose and Miriam. *Mandala*. Berkeley: Shambala Publications, 1972.

Arnheim, Rudolph. *The Power of the Center*. Berkeley: University of California Press, 1982.

Ball, Rouse. *A Short Account of the History of Mathematics*. New York: Dover, 1960.

Billings, Robert. *Power of Form Applied to Geometric Tracery*. London: W. Blackwood and Sons, 1851.

Bouleau, Charles. *The Painter's Secret Geometry*. New York: Harcourt, Brace and World, 1963.

Capra, Fritjof. *The Tao of Physics*. New York: Bantam Books, 1965.

Capra, Fritjof. *The Turning Point*. New York: Bantam Books, 1983.

Colman, Samuel. *Nature's Harmonic Unity*. New York: Putnam's and Sons, 1911.

Cook, Theodore A. *The Curves of Life*. New York: Dover, 1979.

Coxeter, H.S.M. and M. Emmer, R. Penrose and M. L. Teuber. *M. C. Escher: Art and Science*. Amsterdam: North Holland, 1986.

Critchlow, Keith. *Islamic Patterns*. New York: Schocken Books, 1976.

Doczi, Gyorgy. *The Power of Limits*. Boulder: Shambala Press, 1981.

Feininger, Andreas. *The Anatomy of Nature*. New York: Dover, 1956.

Fleming, William. *Arts and Ideas*. New York: Holt, Rinehart and Winston, 1980.

Foy, Sally. *The Grand Design: Form and Color in Animals*. Englewood Cliffs: Prentice Hall, 1982.

Ghyka, Matila. *The Geometry of Art and Life*. New York: Dover, 1977.

Gombrich, Ernst. *The Sense of Order*. Ithaca: Cornell University Press, 1979.

Grunbaum and Shephard. *Tilings and Patterns*. New York: W.H. Freeman and Co., 1990.

Hogben, Lancelot. *Mathematics in the Making*. New York: Doubleday, 1960-61.

Huntley, H.E. *The Divine Proportion*. New York: Dover, 1970.

Kline, Morris. *Mathematics and the Physical World*. New York: Oxford University Press, 1953.

Kline, Morris. *Mathematics in Western Culture*. New York: Oxford University Press, 1953.

Lawler, Robert. *Sacred Geometry*. New York: The Crossroad Publishing Company, 1982.

Loeb, Arthur. *Color and Symmetry*. New York: John Wiley & Sons, 1971.

Mandelbrot, Benoit. *The Fractal Geometry of Nature*. New York: W.H. Freeman and Co., 1983.

Peitgen, Heinz Otto and P.H. Richter. *The Beauty of Fractals*. New York: Springer-Verlag, 1986.

Peitgen, Heinz Otto and Dietmar Saupe. *The Science of Fractal Images*. New York: Springer-Verlag, 1988.

Peterson, Ivars. *Islands of Truth*. New York: W.H. Freeman and Co., 1990.

Peterson, Ivars. *The Mathematical Tourist*. New York: W.H. Freeman and Co., 1988.

Seneschal, Marjorie and George Fleck. *Patterns of Symmetry*. Amherst: University of Massachusetts Press, 1977.

Stevens, Peter S. *Patterns in Nature*. Boston: Little, Brown and Co., 1974.

Thompson, D'Arcy. *On Growth and Form*. Cambridge: University Press, 1969.

Washburn, Dorothy K. and Donald W. Crowe. *Symmetries of Culture Theory and Practice of Plane Pattern Analysis*. Seattle: University of Washington Press, 1988.

Weyl, Hermann. *Symmetry*. Princeton: University Press, 1952.

Williams, Christopher. *Origins of Form*. New York: Visual Communications Books, 1981.

A *Mathematical Symbols*

$\overline{AB}$ — segment AB

$\overleftrightarrow{AB}$ — line AB

$\overrightarrow{AB}$ — ray AB

AB — length of segment AB (designates a real number)

$\triangle ABC$ — triangle ABC

$\angle ABC$ — angle ABC

$\cong$ — is congruent to

$=$ — is equal to

$\approx$ — is approximately equal to

$\sim$ — is similar to

$\perp$ — is perpendicular to

$>$ — is greater than

$\geq$ — is greater than or equal to

$<$ — is less than

$\leq$ — is less than or equal to

$\neq$ — is not equal to

$\pm$ — plus or minus (both operations designated by a single symbol)

$\therefore$ — therefore

$x°$ — x degrees

y' — y minutes (angle measure)

z'' — z seconds (angle measure)

x' — x feet (linear measure)

y'' — y inches (linear measure)

π — pi $\approx$ 3.14 or 22/7 (the ratio of the circumference of a circle to its diameter)

$\emptyset$ — phi $\approx$ 1.61803 (the Golden Ratio)

x_n — x sub n (a way of naming a variable)

x^n — x to the nth power (multiply x by itself n times)

Properties and Proofs *B*

Contents

Properties of Geometric Figures

Polygons

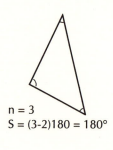 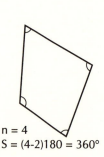 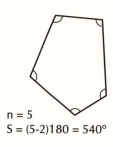

n = 3
S = (3-2)180 = 180°

n = 4
S = (4-2)180 = 360°

n = 5
S = (5-2)180 = 540°

1 The sum, S, of the interior angles of a polygon of n sides is S=(n-2)180.

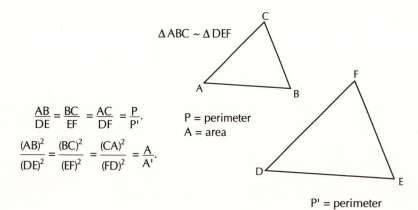

$\triangle ABC \sim \triangle DEF$

$\dfrac{AB}{DE} = \dfrac{BC}{EF} = \dfrac{AC}{DF} = \dfrac{P}{P'}.$

$\dfrac{(AB)^2}{(DE)^2} = \dfrac{(BC)^2}{(EF)^2} = \dfrac{(CA)^2}{(FD)^2} = \dfrac{A}{A'}.$

P = perimeter
A = area

P' = perimeter
A' = area

2 The perimeters of two similar polygons are proportional to the corresponding sides, but their areas are proportional to the squares of the corresponding sides.

Triangles

1 Two triangles are congruent if the following corresponding parts are congruent:

a. one side and two corresponding angles, (ASA, SAA),

b. two sides and the included angle, (SAS),

c. two sides and the right angle opposite one of them, (HL),

d. three sides, (SSS).

a

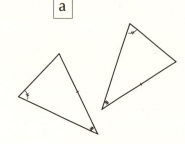

The abbreviations name the corresponding congruent parts: S means side, A means angle, and H stands for hypotenuse. The parts are named in order as one travels around the triangle in either a clockwise or counterclockwise direction.

In each case, the corresponding congruent parts of the triangles are sufficient to insure the congruence of the triangles themselves.

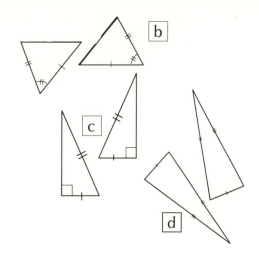

2 Two triangles are similar if:

a. two angles of one triangle are congruent to two angles of the other,

b. an angle of one triangle is congruent to an angle of the other and the corresponding including sides are proportional,

c. three sides of one triangle are in proportion to three sides of the other,

d. three sides of one triangle are parallel to three sides of the other. This is illustrated in example b.

In each case, the listed conditions are sufficient to insure the similarity of the triangles.

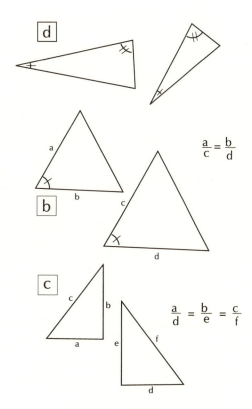

3 The altitudes of a triangle meet in a single point. See altitude in the glossary.

4 The medians of a triangle meet in a point that is two-thirds of the distance from each vertex to the midpoint of the opposite side.

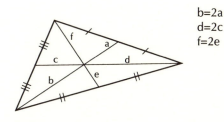

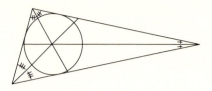

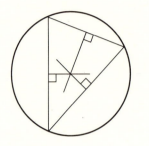

5 The angle bisectors of a triangle meet in a point that is the center of the inscribed circle.

6 The perpendicular bisectors of the sides of a triangle meet in a point that is the center of the circumscribed circle.

$DE = \frac{1}{2}CB$

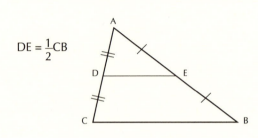

7 The line segment between the midpoints of two sides of a triangle is parallel to the third side and half as long.

$CD = \frac{1}{2}AB$

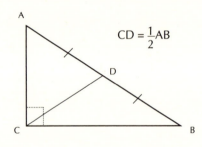

8 The median to the hypotenuse of a right triangle is half as long as the hypotenuse.

$AC = \frac{1}{2}AB$

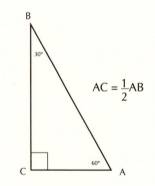

9 In a 30-60-90 Triangle, the side opposite the 30° angle is half the length of the hypotenuse.

$AB = AC\sqrt{2}$
or
$AB = BC\sqrt{2}$

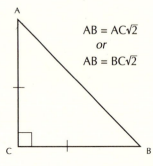

10 In an isosceles Right Triangle, the length of the hypotenuse equals the length of a leg times $\sqrt{2}$.

Quadrilaterals

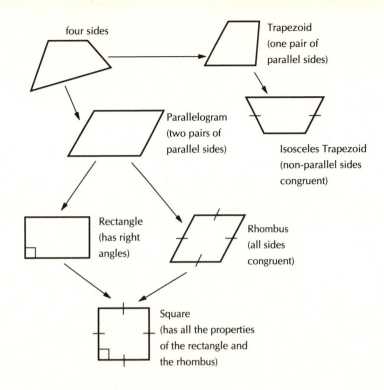

four sides

Trapezoid
(one pair of
parallel sides)

Parallelogram
(two pairs of
parallel sides)

Isosceles Trapezoid
(non-parallel sides
congruent)

Rectangle
(has right
angles)

Rhombus
(all sides
congruent)

Square
(has all the properties
of the rectangle and
the rhombus)

1 In a parallelogram:
   a. opposite sides are congruent,
   b. opposite angles are congruent,
   c. the diagonals bisect each other, and
   d. a diagonal divides the parallelogram into two congruent triangles.

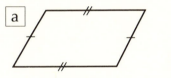

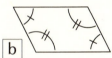

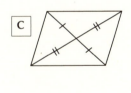

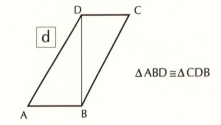

$\triangle ABD \cong \triangle CDB$

2 The diagonals of a rectangle are congruent.

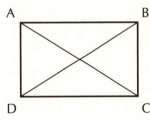

$\overline{AC} \cong \overline{DB}$

3 The diagonals of a rhombus are perpendicular.

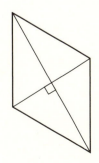

241

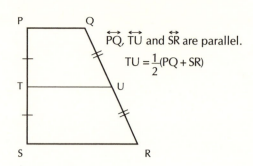

$\overleftrightarrow{PQ}$, $\overleftrightarrow{TU}$ and $\overleftrightarrow{SR}$ are parallel.

$TU = \frac{1}{2}(PQ + SR)$

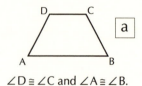

∠D ≅ ∠C and ∠A ≅ ∠B.

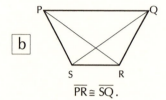

$\overline{PR} \cong \overline{SQ}$.

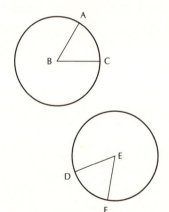

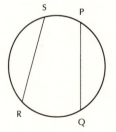

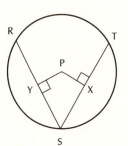

|4| The line segment connecting the midpoints of the non-parallel sides of a trapezoid is parallel to the bases and half as long as the sum of their lengths.

|5| In an isosceles trapezoid:
　　a. the base angles are congruent,
　　b. the diagonals are congruent.

|Circles|

|1| In the same circle or in congruent circles, congruent central angles cut congruent arcs, and conversely.

Assume circles B and E are congruent. If ∠ABC ≅ ∠DEF then arc AC ≅ arc DF, and if arc AC ≅ arc DF then ∠ABC ≅ ∠DEF.

|2| In the same circle or in congruent circles, congruent chords cut congruent arcs, and conversely.

If $\overline{PQ} \cong \overline{RS}$ then arc PQ ≅ arc RS, and if arc PQ ≅ arc RS then $\overline{PQ} \cong \overline{RS}$.

|3| In the same circle or in congruent circles, congruent chords are equidistant from the center.

If $\overline{RS} \cong \overline{ST}$ then $\overline{XP} \cong \overline{PY}$.

4 A line perpendicular to a diameter at one endpoint is tangent to the circle, and conversely.

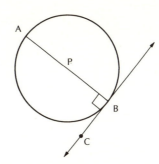

If $\overleftrightarrow{AB} \perp \overleftrightarrow{BC}$ then $\overleftrightarrow{BC}$ is tangent to circle P, and if $\overleftrightarrow{BC}$ is a tangent line, then $\overleftrightarrow{AB} \perp \overleftrightarrow{BC}$.

5 Angles inscribed in the same arc or congruent arcs are congruent.

$\angle A \cong \angle B \cong \angle C \cong \angle D$.

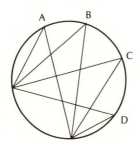

6 Any angle inscribed in a semicircle is a right angle.

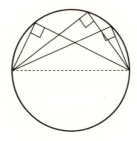

7 A circle, and only one circle, can be circumscribed about or inscribed in any triangle or any regular polygon.

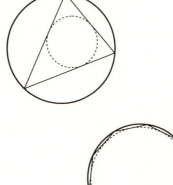

Graphing in Polar Coordinates

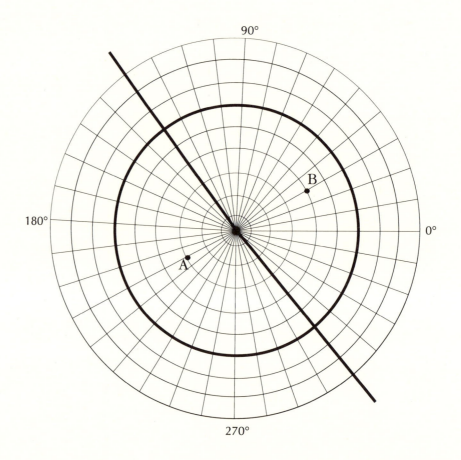

An ordered pair of polar coordinates consists of a value r, the distance from the pole, and a value θ, the measure of the angle in standard position. For positive values of θ, the angle is measured counterclockwise from the 0° vector, and for negative values the angle is measured in a clockwise direction from the 0° vector. This means that (3, 210°) and (3, -150°) both name the same point (A on the diagram). If r is negative, the point is located on the radius vector opposite the named value of θ. For example, (4, 30°) and (-4, 210°) name the same point (B on the diagram).

The graph of an equation of the form r = c, where c is any real number, is a circle. The circle shown on the diagram has as its equation r = 6.

The graph of an equation of the form θ = d, where d represents any number of degrees, is a straight line. θ=130° is illustrated.

Comparative Symmetry Notations

This Text		Standard Crystallography		This Text	Standard Crystallography
Point Groups				*Plane Groups*	
P1	P1'	1	1m	PLP1	p1
P2	P2'	2	2mm	PLP1'	pm
P3	P3'	3	3m	PLGR	pg
P4	P4'	4	4mm	PLP1'GR	cm
P5	P5'	5	5m	PLP2	p2
P6	P6'	6	6mm	PLP2GR	p2gg
				PLP2'	p2mm
Line Groups				PLP2R	p2mg
				PLP2RR	c2mm
LT		t		PLP4	p4
LGR		tg		PLP4'	p4mm
LRH		mt		PLP4RR	p4gm
LRV		tm		PLP3	p3
LRR		t2mm		PLP3R	p31m
LRo		t2		PLP3'	p3m1
LRRo		t2mg		PLP6	p6
				PLP6'	p6mm

Hints for Analysis of Plane Patterns

Determine whether the pattern is periodic or nonperiodic. If nonperiodic the pattern cannot be analyzed with relation to the plane groups of symmetry. If the pattern is a periodic one, the following are hints for recognizing the unit:

1. Find a motif that is repeated in the pattern. Remember that a unit is created by symmetry operations performed on a motif.

2. Check to see if the motif is asymmetric or has bilateral symmetry. This will help to determine the type of point group that is present within the unit.

3. Lay a piece of tracing paper over the motif and outline the boundaries. Slide

the tracing around on the pattern and note what kind of moves are needed to get the motif to correspond to the same image elsewhere. Are the moves straight translations? Must the image be rotated? Do you have to flip the paper over because the image is reflected?

4. Try to determine how the motif is used to create a unit. Is it used singly? Is there a center of rotation? How many times is the motif rotated? Answers to these questions are directly related to particular plane groups. Check to see which use point groups of order 1, 2, 3, 4 or 6.

5. Try to find implied lines that run through the same points on units as they are translated. If lines of translation

intersect at right angles, the unit must be a rectangle. If not, try to determine if the unit is a parallelogram or a hexagon. Be aware of which plane groups have which kinds of regions, and that parts of the unit may overlap into other units.

6. Spaces between units effect the design, but do not figure into the determination of which plane group is involved. For example, a glide reflection may be a full glide or a half glide.

These hints refer only to symmetry as it relates to structure of the pattern, not to color, which may or may not be symmetrically incorporated into the pattern.

Properties of Penrose Tiles

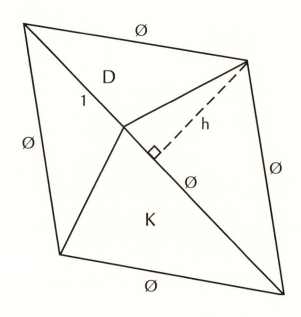

The ratio of the area of a kite to that of a dart is Ø.

Proof: Let K be a kite and D be a dart. If K and D are fitted together to form a rhombus, then h is the altitude of one of the triangles making up K and also of one of the triangles making up D. We can find the areas of K and D by doubling the areas of those triangles. The areas are:
A(K) = 2(1/2)Øh = Øh
and A(D) = 2(1/2)h = h.

Then $\dfrac{A(K)}{A(D)} = \dfrac{Øh}{h} = Ø.$

The ratio of the area of a "fat" rhombus to that of a "skinny" one is Ø.

Proof: Let F be a "fat" rhombus and S be a "skinny" one. F and S can be overlapped as in the diagram so that h is the altitude of F and also of one of the Golden Triangles that make up S. The areas of F and K are as follows:
A(F) = Øh
and A(S) = 2(1/2)1h = h.

Then $\dfrac{A(F)}{A(S)} = \dfrac{Øh}{h} = Ø$.

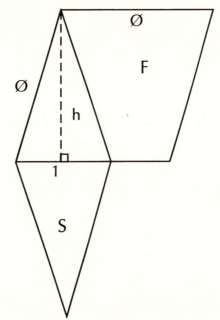

Arithmetic of Complex Numbers

Any pair of complex numbers may be added, subtracted, multiplied or divided to produce another complex number. Let a + bi and c + di be any complex numbers.

To add: (a + bi) + (c + di) = (a + c) + (b + d)i.

To subtract: (a + bi) - (c + di) = (a - c) + (b - d)i.

To multiply use FOIL (add the products of the First, Outside, Inside and Last terms):

$$(a + bi)(c + di) = ac + adi + bci + bdi^2$$
$$= ac + (ad + bc)i - bd \quad (\text{since } i^2 = -1)$$
$$= (ac - bd) + (ad + bc)i.$$

Definition: the *conjugate* of a complex number a + bi is the number a - bi.
To divide, write the division problem as a fraction and multiply both the numerator and the denominator by the conjugate of the denominator.

$$(a+bi) \div (c+di) = \frac{a+bi}{c+di} = \frac{a+bi}{c+di} \cdot \frac{c-di}{c-di} = \frac{ac - adi + bci - bdi^2}{c^2 - cdi + cdi - d^2i^2} = \frac{(ac + bd) + (bc - ad)i}{c^2 + d^2}$$

$$= \left(\frac{ac + bd}{c^2 + d^2}\right) + \left(\frac{bc - ad}{c^2 + d^2}\right)i.$$

C *Plane Figure Geometric Formulas*

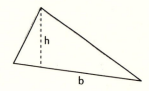

Triangle

Area: $A = \frac{1}{2}bh$ where b is the measure of any side and h is the measure of the altitude to that side.

$A = \sqrt{s(s-a)(s-b)(s-c)}$ where a, b and c are the lengths of the sides and $s = \frac{1}{2}(a+b+c)$.

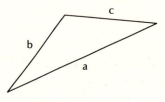

Right Triangle

Pythagorean Theorem: $a^2 + b^2 = c^2$ where a and b are the measures of the legs and c is the measure of the hypotenuse.

Square

Area: $A = s^2$

Perimeter: $P = 4s$ where s is the measure of a side.

Rectangle

Area: $A = lw$

Perimeter: $P = 2l + 2w$ where l is the measure of the length and w is the measure of the width.

Parallelogram

Area: A = bh where *b* is the measure of a side and *b* is the measure of an altitude to that side.

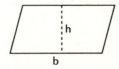

Rhombus

Area: $A = \frac{1}{2}ab$ where *a* and *b* are the lengths of the diagonals.

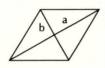

Trapezoid

Area: $A = \frac{1}{2}h(a + b)$ where *a* and *b* are the measures of the parallel sides and *b* is the distance between those sides.

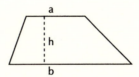

Circle

Area: $A = \pi r^2$

Circumference: $C = \pi d = 2\pi r$ where *r* and *d* are the radius and diameter, respectively, and .π may be approximated by 3.14 or 22/7.

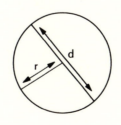

Regular Polygon

Area: $A = \frac{1}{2}asn$

Perimeter: P = sn where *a* is the measure of the apothem, *s* is the measure of a side, and *n* is the number of sides.

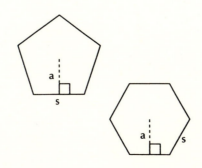

D *Art Materials &Techniques*

Contents

Approaching an Art Project

If doing artwork is new to you, the following steps may help you on your journey.

1 Decide upon the subject matter of the work, which might be still life, portraiture, landscape or pure design. What elements do you find most interesting about this particular subject matter? What is most important to you about this subject? What qualities do you want to emphasize?

2 Choose the size, shape and proportions of your two dimensional support, which might be canvas, paper, or board. Do this carefully. The outer proportions will determine how successfully you can subdivide the interior surface-space. Certain proportions create dynamic and harmonic divisions, while others do not subdivide well. (See Book 1, *Universal Patterns.*)

Be concerned with size. Some ideas are more intimate and require smaller dimensions. Some ideas are large enough for murals. In some ways, it is analogous to the differences between a poem, a short story or a novel.

3 Work out the arrangement and placement of the large shapes on an armature. Do not consider the specific details at this time. They will come later. Do a group of quick sketches for potential arrangements of forms. Choose at least three ideas from the sketches. Refine them in a slightly larger size. This might be the place to experiment with different materials. Keep in mind the proportions of your original format and consider the following:

a. Will you be working with a color structure or only with black and white?

b. Are you working only with a strong dark and light contrast?

c. Are you going to have a dark, a light and two or three other value changes?

d. Are you going to work with only 3 - 5 light values?

e. Are you going to work with only 3 - 5 dark values?

In your studies, either find your values by your choice of papers, or create them by covering sheets of paper with paint, colored pencils, markers or inks in the values of your choice. Then cut them into the desired shapes and glue onto the surface. Consider the other options of texture and pattern.

4 Once the values and placement of key shapes are established, block them onto your final composition.

5 On top of this understructure, you will be working in your basic colors and adding your details. Decide where your center of interest will be and work to bring all the color, movement and rhythms to that place. Pay attention to detail and craftsmanship.

6 Bon voyage!

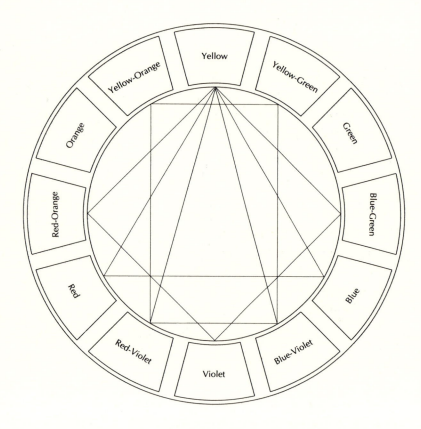

Labels on wheel: Yellow, Yellow-Green, Green, Blue-Green, Blue, Blue-Violet, Violet, Red-Violet, Red, Red-Orange, Orange, Yellow-Orange

Tints are obtained by mixing the color with white, while *tones* are gotten by mixing gray into a color, and *shades* are obtained by mixing in black.

In this particular color wheel, the geometric figures in the center circle provide different color structures. They can be placed anywhere within the circle and the vertices will indicate which colors will work together. The two triangles give a triadic color structure (one made up of three colors) while the rectangles give tetradic structures (made up of four colors).

Do not always rely on given systems, however. Be experimental in the mixing and placing of colors.

Color Harmonies

In order for a color scheme to be harmonious, a few basic guidelines could be followed in limiting the number of colors used. This can be done by using one of the following schemes:

Monochromatic

Uses one color and the light and dark variations thereof.

Complementary

Uses a pair of colors directly opposite one another with light and dark variations thereof.

Analogous

Uses three adjacent colors with light and dark variations thereof.

Triad

Uses three colors that form the vertices of either an isosceles or equilateral triangle with the light and dark variations thereof.

Tetrad

Uses four colors that form the vertices of a rectangle (including the square) with the light and dark variations thereof.

Color Wheel

The color wheel is based on a twelve tone structure. Yellow, blue and red are the three *primary* colors. Theoretically, all the other colors can be mixed from only these three. It is best, however, when purchasing pigments to include white, black, and violets to gain the maximum amount of flexibility in mixing color.

A *secondary* color is obtained by mixing a pair of primary colors. The secondary colors are orange, green and violet.

A mixture of a primary and a secondary color yields one of the *tertiary* colors, which are the remaining ones on the wheel.

Variation on the Birren Color Triangle

Color harmonies can alsobe constructed by looking at the Birren color triangle and choosing one of the following:

> Color (your choice) to tone to
>   gray
> Color to tint to white
> Color to shade to black
> Black to gray to white
> Tint to tone to shade
> Shade to tone to white
> Color to tint to white to gray to
>   black
> Tint to shade to tone to gray

Tint: a color with white added.

Tone: a color with gray* added.
    *Gray: black and white mixed together or a pair of complementary colors mixed together.

Shade: a color with black added.

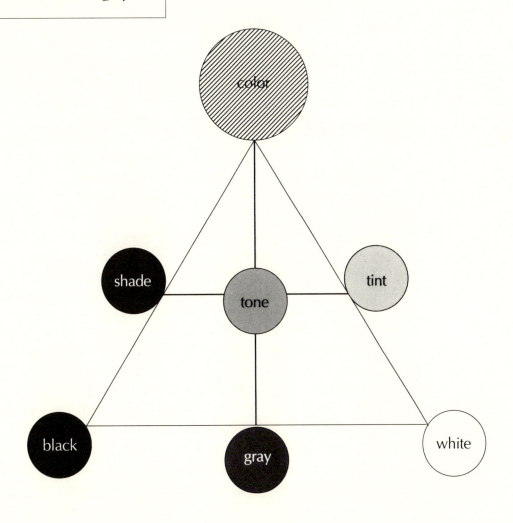

Basic Materials

The following materials are suggested if you are just beginning your investigation of art. This list is by no means complete as the modern market abounds in art supplies of every description. It is, rather, a list of basic supplies, relatively easy to use, offered with the hope that your first attempts at artistic endeavors will be successful and will encourage you to continue.

Adhesives

See Adhesive Chart on page 256.

Brushes

Brushes are constructed for use with different media. There is a synthetic type designed for acrylic polymer paints. Which type will depend upon the consistency of the paint used and the effect desired. Have at least three different sizes, ranging from fine for detail work to 1/2" for fill-in work.

Construction Tools

Metal straightedge

Compass with opening wide enough for constructing good sized circles

Scissors

Utility knife with extra blades

T-square and triangles

Fabric Crayons

These can be used to draw original designs on paper which can then be heat transferred by iron to fabric. The design is reversed when transferred.

Inks

These come in permanent, nonpermanent, transparent, opaque and metallic types. They can be used in combination with other materials.

Markers

These come in a great many varieties from water base to permanent (this only refers to the markers being waterproof, not lightfast); from transparent to opaque; from fluorescent to metallic. Test before buying in quantity.

Paints

Gouache—water base opaque paint with an essentially matte finish—not necessarily permanent.

Acrylic—water base but permanent when dry.

Include: (2 oz. tubes or jars)
white   yellow   black   blue
red   violet or purple
medium (matte or gloss type)

Traditional transparent watercolor—water based paint which is not permanent. Allows the white of the paper to show through as a visual element. These paints can be used transparently or opaquely depending upon the desired effect. Good coverage may require between three and five coats. Brushes should be kept in water to prevent the paint from hardening and ruining them. This can happen very quickly since the paint is fast drying.

Paper

See Paper Chart on page 255.

Pencils

Graphite pencils

H = hard   B = soft The number and the letter on the side of the pencil indicates degree of hardness or softness. HB is a middle range.

Colored pencils

These come in a variety of types. Some can be erased, some are water soluble and can be used like watercolor, some are oil base. Buy a good quality. They can be purchased individually or in sets, in hard or soft types, and in stick form. They are blended by drawing on top of one another. To obtain clean, precise lines and to reduce smudging, it would be best to avoid crayons, oil pastels, pastels, and charcoal.

Papers and Their Uses

TYPE	WEIGHT	CHARACTERISTICS	TEXTURE & COLOR	USES
Albanene	Light	100% rag, transparent easily erasable, will not yellow or crack	White, smooth	Tracing
Bristol board	Light to heavy	Tough, many thicknesses	Very smooth, white	Acrylic studies, sculpture
Bond	Light	Somewhat fragile	White, smooth	Pencil studies
Carbon transfer	Very light	Carbon on one side	Several colors	Transfer
Construction	Medium	Inexpensive, fades easily	Full range of color	All-purpose, temporary
Graphite	Light	One side coated with graphite	Gray	Transfers to board & paper
Illustration board	Light to heavy	Lies flat, consistent working surface	White, grays, blacks smooth and rough surfaces	Polyhedra models, acrylic studies
Craft	Medium	Strong	Brown	All purpose
Metal foil	Light	Paper backed with foil	Varied color, shiny and matte	Sculpture
Oak tag	Heavy	Strong, folds without cracking	Full range of colors, smooth	Polyhedra models, paper sculpture
Press-apply	Light	Adhesive backed	Smooth, intense rainbow colors	Surface decoration
Scorasculpture (Dennison Co.)	Medium	Easily scored, very workable	Smooth, white	Polyhedra models, paper sculpture
Tissue	Very light	Translucent	Full range of color, plain or variegated	Surface decoration
Velour	Medium	Flocked, velvet-like surface	Full range of color	Surface decoration

Adhesives & Their Uses

A	Contact Cement	F	Resin Glue
B	Epoxy Cement	G	Rubber Cement Glue
C	Epoxy Metal Cement	H	Waterproof Glue
D	Plastic Resin Urea Base	I	White Glue
E	Polymer Medium (Acrylic Emulsion)	N	Nonporous Material
		P	Porous Material

MATERIALS TO BE GLUED	A	B	C	D	E	F	G	H	I	N/P
China and glass, ceramic tiles	x								x	
Collagraph items for printmaking				x			x			x
Cloth with wood					x	x		x		x
Leather and wood		x			x	x		x		x
Metal to metal	x	x	x					x		
Metal to wood	x	x	x			x		x	x	
Paper and cloth					x	x		x		x
Paper to paper and cardboard					x	x	x	x		x
Plastic foams (Styrofoam)		x		x	x	x		x	x	x
Plastics and Vinyls				x	x	x			x	
Plastic to wood		x		x		x			x	
Rubber							x	x	x	
Rubber to wood or metal	x	x	x						x	
Stone & concrete to other items	x								x	
Sand, stones, beads				x	x	x		x	x	
Wood outdoors				x			x			x
Wood to wood				x		x		x	x	x

General Comments:

Labels should be checked for proper use of the product.

When gluing dissimilar materials, their porosity should be determined. When combining two types of materials, the type recommended for the non-porous material should be used.

Duco Cement or LePage's Household Cement are good all-purpose adhesives. LePage's Glue is good for general paper work.

Shoe Patch, purchased at a sporting goods store, is a very strong adhesive that might be tried when others fail.

Polymer medium is useful for preserving, glazing and sealing. It comes in both matte and gloss finish.

With the exception of polymer medium, available at art supply stores, all other adhesives can be found in hardware stores.

Adhesives should be tested on a small sample first.

Paper Adhesives	Characteristics	Uses
Glue stick	Very tacky, very strong	Adhering tabs and small areas
Metylan cellulose paste	Inexpensive, dissolves in water	Paper maché, paper, cardboard
Mucillage	Amber colored, tacky, syrupy	Paper
Polyvinyl acetate (Elmer's, Sobo, Duratite, etc.)	White, dries clear, strong	Decoupage, paper maché
Plastic glaze (Modge Podge, Art Podge)	White, dries clear, strong	Protecting paper surfaces
Rubber Cement	Dries quickly, excess easily rubbed away	One coat temporary, two coats permanent

E Grid Templates

Contents

Square Dot Array

Isometric Dot Array

Square Grid

Isometric Grid

Golden Grid Units are Golden Rectangles — Ratio of lengths of sides: 1 to 1.618 (approx.)

Golden Section Grid A Units are squares and Golden Rectangles — Ratios: 1 to 1 and 1 to .618 (approx.)

Golden Section Grid B Units are squares and Golden Rectangles — Ratios: 1 to 1, .618 to .618, and 1 to .618 (approx.)

Ø + 1 Grid

Units are Ø + 1 Rectangles — Ratio of
lengths of sides: 1 to 2.618 (approx.)

$\boxed{\sqrt{2}\ \text{Grid}}$ Units are $\sqrt{2}$ Rectangles — Ratio of lengths of sides: 1 to 1.414 (approx.)

$\boxed{\sqrt{3}\ \text{Grid}}$ Units are $\sqrt{3}$ Rectangles — Ratio of lengths of sides: 1 to 1.732 (approx.)

$\sqrt{4}$ Grid Units are $\sqrt{4}$ Rectangles — Ratio of lengths of sides: 1 to 2

$\sqrt{5}$ Grid Units are $\sqrt{5}$ Rectangles — Ratio of lengths of sides: 1 to 2.236 (approx.)

"Fat" Penrose Rhombus Grid

"Skinny" Penrose Rhombus Grid

F *Answers to Selected Problems*

It is highly recommended that you do each of the constructions in a given chapter before attempting the problems, some of which require knowledge of the constructions. For quite a few of the problems, there are many "right" answers. Therefore, much of this section consists of hints or directions on how to start.

Chapter 1

1. A (5/4, 0), B (3, 2), C (-5/3, 1), D (-2, -1), E (0, -3), F (3, -1). Any fractional coordinates are approximate.
2. First coordinates are given rounded to the nearest whole number, and second coordinates are given rounded to the nearest 10°. Clockwise from the upper right: (7, 30°), (7, 160°), (7, 210°), (9, 270°), (6, 340°).
4. Follow directions in Construction 4.
5. Follow directions in Construction 3.
7. Follow directions in Construction 2.

Chapter 2

5. See Appendix B for information on analyzing a pattern with regard to plane group symmetry.

Chapter 3

2. 4/5" (The diameter of each circle is 8/5 — the length of the side of the square divided by 5, and the radius of each would be 1/2 of that.)
3. Use central angles of approximately 17.1°.
4. See pp. 104 - 105.
6. Follow directions in Construction 13.
7. See Fig. 3.31.

Chapter 4

1. 3-8-24, 3-7-42, 3-9-18, 3-10-15, 4-5-20, 5-5-10.
5. See p. 151.

Chapter 5

2. The eighth iteration is $f(1.9348 \times 10^{24}) = 3.7434 \times 10^{48}$.
3. The tenth iteration is $f(36 - 18i) = 40 - 20i$.
6. $4 + i$, $3 + 0i$ or 3, $2 - 5i$, $0 - 6i$ or $-6i$.

Chapter 6

1. Follow directions in Construction 18.
2. Any of the techniques described in the chapter could be used for this problem.
7. For the bubble follow directions in Construction 19. See Fig. 6.16 for irregular distortions of the grid.
8. Follow directions in Construction 20 after first drawing an irregular prism on an isometric grid.

Glossary

A

acute angle An angle that measures between 0° and 90°.

acute angles

acute triangle A triangle having three acute angles.

acute triangles

adjacent Next to.

algorithm A mathematical rule.

altitude *In a triangle:* the line segment from a vertex perpendicular to the opposite side.

Every triangle has three altitudes which might not all be in the interior of the triangle.

In an acute triangle all three altitudes lie in the interior of the triangle.

altitudes: $\overline{AF}, \overline{BD}, \overline{CE}$

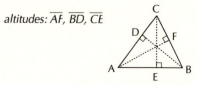

In a right triangle the legs are also altitudes

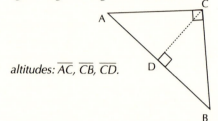

altitudes: $\overline{AC}, \overline{CB}, \overline{CD}.$

In an obtuse triangle two of the altitudes lie in the exterior of the triangle.

altitudes: $\overline{CF}, \overline{BE}, \overline{AD}.$

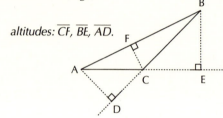

In a parallelogram: the altitude is the distance between two parallel sides. Any line segment joining two sides and perpendicular to both is *an* altitude.

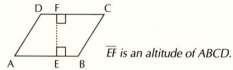

$\overline{EF}$ *is an altitude of ABCD.*

analogous color scheme A color scheme that uses hues that are next to each other on the color wheel. Light and dark values of the hues may be included.

anamorphism The transformation of images or natural forms in evolution.

angle The figure formed by two rays having a common endpoint. The common endpoint is called the vertex of the angle, and the rays are called the sides of the angle.

∠ABC names the angle whose vertex is B such that A is a point on one side and C is a point on the other. This angle could also be named ∠CBA. The vertex B must be between the other two points named.

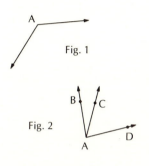

Fig. 1

Fig. 2

If there is no ambiguity, an angle may be named only by its vertex. ∠A makes sense in Fig. 1 but not in Fig. 2 since there are three angles at A, namely ∠BAC, ∠CAD , a n d ∠BAD.

arc The figure formed by two points and the portion of a circle between those two points.

The two points are called the endpoints of the arc, and an arc is named by its endpoints.

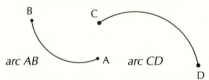

arc AB   arc CD

If two points lie on a circle, they determine two different arcs. To avoid confusion, the arcs may be named by the endpoints and any other point on the arc. The third point is listed between the endpoints. In Fig. 1, A and B are endpoints of a diameter. Therefore, arc ACB and arc ADB are both semicircles. In Fig. 2, A and B are not endpoints of a diameter. Arc ACB is called the minor arc (the smaller one) and arc ADB is called the major arc (the larger one).

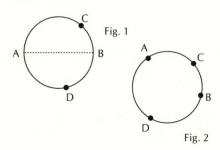

Fig. 1

Fig. 2

area The measure in square units of any bounded portion of a plane.

axis A line used as a reference.

In the Cartesian coordinate plane all points are located with reference to two perpendicular axes: the x axis and the y axis.

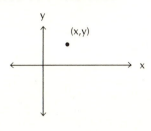

In the complex plane a number, a + bi, is located so that a corresponds to a coordinate on the horizontal, or real, axis and b corresponds to a coordinate on the vertical, or imaginary, axis.

*An axis of symmetry is a line which divides the plane into two halves which are mirror images of each other.
Also called an axis of reflection or a mirror.*

Axis of reflection See axis.

Axis of symmetry See axis.

B

base angles *In an isosceles triangle:* the two congruent angles.

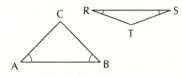

∠A and ∠B are the base angles of ∆ABC, and ∠R and ∠S are the base angles of ∆RST.

In a trapezoid: a pair of angles with a base as a common side. Each trapezoid has two pairs of base angles.

See diagram next page.

In *ABCD*, ∠ *A* and ∠ *B* are one pair of base angles, and ∠ *C* and ∠ *D* are the other pair.

base group Any one of the point groups developed from rotation alone.

The base groups are named with an upper case P followed by a numeral which indicates the number of times the motif is repeated in one complete rotation.

base group P2

bisect To divide into two congruent figures.

Line l bisects $\overline{PQ}$.

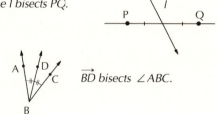

$\overrightarrow{BD}$ bisects ∠ ABC.

bisector The line or ray that divides a segment or angle into congruent figures.

bisector, perpendicular A line that passes through the midpoint of a segment and is at right angles to the segment.

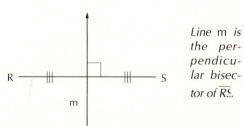

Line m is the perpendicular bisector of $\overline{RS}$.

C

Cartesian coordinate plane A rectangular system for locating points in a plane by using a pair of perpendicular axes: the horizontal line is the x axis and the vertical line is the y axis.

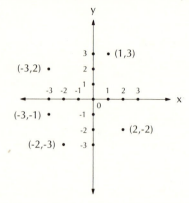

A point is located by an ordered pair of numbers (x,y) which places it with respect to both axes.

center The point from which all points on a circle are equidistant.

P is the center of the circle.

central angle The angle formed by two radii of a circle.

In circle P, ∠ APB is a central angle.

chord A line segment whose endpoints lie on a circle.

In circle P, $\overline{RS}$ is a chord.

circle The set of all points in a plane equidistant from a given point.

Although it is not part of the circle, a circle is named by its center. Circle Q is a circle whose center is Q.

circumference The distance around a circle.

The circumference of a circle is given by the formula C = πd, where d is the diameter of the circle.

coefficient A factor in a product generally taken to mean a number multiplied by a variable quantity.

In the expressions 2x, -3a²b, -pq: 2 is the coefficient of x, -3 is the coefficient of a²b, and -1 is the coefficient of pq.

collinear Lying on the same line.

In the figure, points A, B and C are collinear.

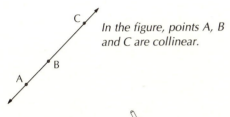

compass An instrument used for drawing circles and arcs .

complementary color scheme A color scheme that uses two hues that are directly opposite each other on the color wheel. It can include light and dark values of the hues.

complex number A number of the form a + bi, where a and b are real numbers and i = $\sqrt{-1}$.

eg. 3 + 2i, i, -1 - 6i, and 0 are all complex numbers. Note: the reals are also complex since any real number can be expressed as a + 0i. If a is 0, the number is said to be purely imaginary.

complex plane A rectangular system composed of a vertical (imaginary) axis and a horizontal (real) axis that allows for the plotting of individual complex numbers.

concave polygon A polygon that has at least one interior angle that measures more than 180°.

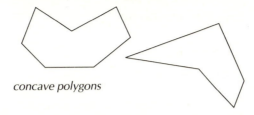

concave polygons

concentric Having the same center.

The two circles with center Q are concentric circles.

concurrent Occurring at the same time.

congruent Having the same size and shape.

consecutive In order or in sequence.

e.g. 2, 3, 4 are consecutive integers. 5 and 7 are consecutive odd integers.

In a polygon, consecutive sides or consecutive vertices are those that appear in order as you travel around the polygon in either direction, starting at any vertex.

convex polygon A polygon whose interior angles each measure less than 180°.

convex polygons

cube *As a noun:* A polyhedron having six square faces.

As a verb: To raise to the third power.
eg. 4³ = 4 cubed = 4x4x4 = 64.

273

curve A continuous set of points having only one dimension, namely length.
Note: by this definition, a straight line is a curve.

D

decagon A polygon having ten sides.

decagons

deduction A method of reasoning that allows specific conclusions to be drawn from general truths. This is the form of reasoning used in mathematical proof.

degree The unit of measure for angles.
There are 360° in a circle.

denominator In a fraction, the number that is written below the line.

diagonal A line segment that joins the nonconsecutive vertices of a polygon.
The number of diagonals a polygon has is dependent on the number of sides.

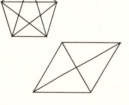

Each of the line segments in the interior of the figures is a diagonal.

diameter A chord that passes through the center of a circle.

A circle has infinitely many diameters—all the same length. The measure of any one of them is called the diameter of the circle. $\overline{QR}$ is a diameter of circle P. QR is the diameter of circle P. A diameter cuts the circle into two semicircles.

dihedral tiling A tiling that has only two prototiles.

distance *Between points:* the length of the line segment joining the points.

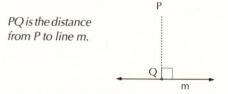

AB is the distance from A to B.

From a point to a line: the length of the line segment from the point to the line and perpendicular to the line.

PQ is the distance from P to line m.

Between parallel lines: the length of any segment having one endpoint on each line and perpendicular to both.

ST is the distance from line l to line m.

Divine Proportion The proportion derived from the division of a line segment into two segments such that the ratio of the whole segment to the longer part is the same as the ratio of the longer part to the shorter part.

$$\frac{AB}{AC} = \frac{AC}{CB}$$

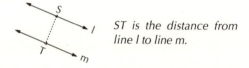

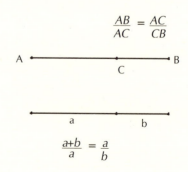

$$\frac{a+b}{a} = \frac{a}{b}$$

dodecagon   A polygon having twelve sides.

dodecagons

Dynamic Parallelogram   Any parallelogram whose sides are in the same ratio as one of the Dynamic Rectangles.

Dynamic Rectangles   A family of rectangles derived from the square through the use of diagonals. If the width is 1, then the lengths of the rectangles are given by square roots of the natural numbers.

The Dynamic Rectangles are sometimes called the Euclidean Series, sometimes root rectangles.

E

edge-to-edge tiling   A tiling in which the tiles share either a side and two vertices or only a vertex.

edge-to-edge tiling

equiangular   Having angles of the same measure.

equiangular triangle

equiangular quadrilateral

equilateral   Having sides of the same measure.

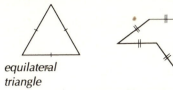

equilateral triangle

equilateral hexagon

escribe   To draw around.

Euclidean Series   See Dynamic Rectangles.

even number   A whole number that is divisible by two.

eg. 2, 4, 6, 8, 10, . . .

If n is any whole number, 2n will always designate an even number.

exponent   A number that indicates the number of times a quantity is to be used as a factor.

In the expression a^4, 4 is the exponent.
$a^4 = (a)(a)(a)(a).$

exterior *(of an angle or a closed curve)* The set of all points in the plane that do not lie on the figure or in the interior of the figure.

The shaded portion of the plane represents the exterior of the figure.

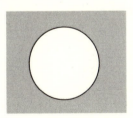

exterior angle of a polygon   An angle formed by a side and the extension of an adjacent side in a polygon.

Note: there are two exterior angles at each vertex.

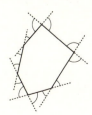

F

factor A number that divides another number without a remainder.
eg. 3 is a factor of 12, since 3 divides 12 4 times without a remainder.

Fibonacci sequence The summation sequence whose terms are: 1, 1, 2, 3, 5, 8, 13, 21, 34, 55, . . .

fractal A curve that exhibits self similarity at all levels of magnification.

function A set of ordered pairs in which no first element is paired with more than one second element.

G

generation In fractal geometry, the figure determined by a particular number of iterations of the rule.

generator In fractal geometry, the figure that results from one application of the rule.
The generator is a second generation curve.

glide reflection The symmetry operation that combines translation and reflection.

gnomon Any figure that when added to a given figure results in a figure similar to the original.

In each figure, the shaded portion is a gnomon for the unshaded triangle.

Golden Cut See Golden Section.

Golden Mean See Golden Section.

Golden Ratio $\frac{1+\sqrt{5}}{2}$, which is the ratio of approximately 1.61803 to 1, derived from the Divine Proportion.
The Golden Ratio is known by the Greek letter, Ø.

Golden Rectangle Any rectangle whose sides are in the ratio of Ø to 1, where, to five decimal places, Ø=1.61803.

Golden Section The point on a line segment that divides the segment into two segments whose lengths are in the Golden Ratio.

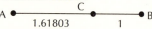

C is the Golden Section of $\overline{AB}$.

Golden Sequence The sequence whose terms are the successive powers of Ø; Ø, Ø², Ø³, Ø⁴, Ø⁵, ...

Golden Spiral The logarithmic spiral formed within the pattern of whirling squares in a Golden Rectangle.

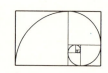

Golden Triangle Any isosceles triangle in which the ratio of a leg to the base is Ø to 1, where, to five decimal places, Ø=1.61803.

Also called the Triangle of the Pentalpha or the Sublime Triangle.

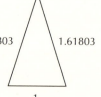

grid A repeating pattern superimposed on the plane consisting of lines and/or circles.

grid, dynamic A grid whose units are

composed of Dynamic Rectangles or Dynamic Parallelograms.

grid, harmonic A grid whose units contain Ø proportions. The units may be triangles, rectangles or parallelograms.

grid, polar A grid consisting of concentric circles with radius vectors emanating from the center, or pole.

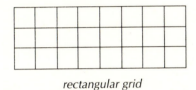

polar grid

grid, rectangular A grid whose units are rectangles.

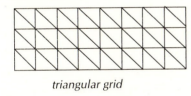

rectangular grid

grid, triangular A grid whose units are triangles.

triangular grid

H

half plane See side.

Harmonic Triangle
The right triangle derived from the radius and apothem of a regular pentagon.

heptagon See septagon.

hexagon A polygon having six sides.

hexagons

hue The name of a particular color.

hypotenuse In a right triangle, the side opposite the right angle.

In △ABC, $\overline{AB}$ is the hypotenuse.

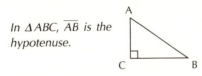

hypothesis A conjecture that requires testing for verification.

I

icosagon A polygon having 20 sides.

icosagon

included angle The angle formed by two specified adjacent sides of a polygon.

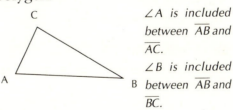

∠A is included between $\overline{AB}$ and $\overline{AC}$.
∠B is included between $\overline{AB}$ and $\overline{BC}$.
∠C is included between $\overline{AC}$ and $\overline{BC}$.

included side In a polygon, the common side between two specified angles.

$\overline{AC}$ is included between ∠A and ∠C.
$\overline{AB}$ is included between ∠A and ∠B.
$\overline{BC}$ is included between ∠B and ∠C.

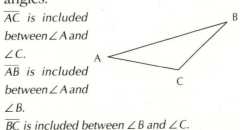

277

induction See inductive reasoning.

inductive reasoning A method of reasoning that allows a generalization to be made from observing specific cases.

Inductive reasoning can be likened to making educated guesses, and, unlike deductive reasoning, can lead to a false conclusion. This is the form of reasoning used in the Scientific Method.

initiator In fractal geometry, the original figure to which the rule is applied (the first generation curve).

inscribed angle An angle whose vertex lies on a circle and each of whose sides intersects the circle.

∠CAB is inscribed in circle P.

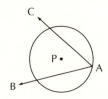

inscribed circle A circle lying in the interior of a polygon such that each side of the polygon is tangent to the circle.

Circle P is inscribed in AB-CDEF.

inscribed polygon A polygon whose vertices lie on a circle.

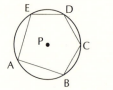

ABCDE is inscribed in circle P.

integer Any member of the following infinite set of numbers:

. . . -3, -2, -1, 0, 1, 2, 3, . . .

interior *of an angle* The set of all points not on an angle having the property that if any two are joined by a line segment, the segment will not

intersect the angle.

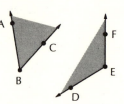

The shaded portions represent the interiors of angles ABC and DEF, respectively.

interior *of a closed curve* The set of all points in the plane that are enclosed by the curve.

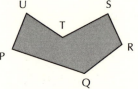

The shaded portion represents the interior of polygon PQRSTU.

interior angle *of a polygon* An angle formed by adjacent sides of a polygon.

The number of interior angles is the same as the number of sides or vertices.

intersect To have at least one point in common.

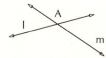

Lines l and m intersect at point A, and A is called the point of intersection.

irrational number Any number that cannot be expressed as the ratio of two integers.

isometric Pertaining to grids consisting of equilateral triangles.

isosceles trapezoid A trapezoid in which the nonparallel sides are congruent.

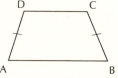

The non-parallel sides are called the legs of the trapezoid.

ABCD is an isosceles trapezoid with legs $\overline{AD}$ and $\overline{BC}$.

isosceles triangle A triangle having a pair of congruent sides.

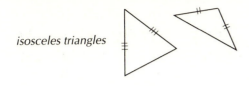

isosceles triangles

iteration A repetitive procedure generally linked to a mathematical function wherein the previous output becomes the input.

L

leg(s) *In a right triangle:* the sides that form the right angle.

In △ABC, $\overline{AC}$ and $\overline{CB}$ are the legs.

In an isosceles triangle: the congruent sides.

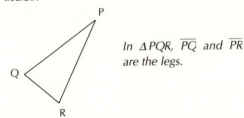

In △PQR, $\overline{PQ}$ and $\overline{PR}$ are the legs.

In an isosceles trapezoid: the nonparallel sides.

In LMNO, $\overline{LO}$ and $\overline{MN}$ are the legs.

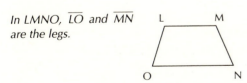

line An undefined term in geometry. It has the properties of infinite length, continuity, and no width. It is a straight curve.

A line is named by any two of its points or by a lower case letter.

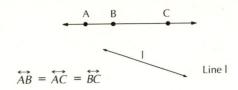

$$\overleftrightarrow{AB} = \overleftrightarrow{AC} = \overleftrightarrow{BC}$$

line group A pattern resulting from translating a unit in a single direction. The unit is developed by manipulating a motif with one or more symmetry operations.

There are seven line groups.

line segment The figure formed by two points on a line and all the points in between those two points.

The two points are called the endpoints of the segment, and a line segment is named by its endpoints.

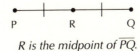

$$\overline{AB} = \overline{BA}$$

M

median A line segment that joins a vertex of a triangle to the midpoint of the opposite side.

Every triangle has three medians.
medians $\overline{AF}$, $\overline{CE}$, and $\overline{BD}$.

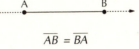

midpoint The point that divides a line segment into two congruent segments.

R is the midpoint of $\overline{PQ}$.

minute One-sixtieth of a degree in angle measure.

46°17' reads 46 degrees, 17 minutes. 60' = 1°.

mirror See axis

moire A pattern consisting of two or more superimposed grids.

monochromatic color scheme A color scheme that uses one hue and the light and dark values thereof.

monohedral tiling A tiling that has a single prototile.

morphology The study of form.

N

natural numbers The counting numbers.

The set of natural numbers is named by the letter N. N = {1,2,3, ...}

net A bounded portion of a grid or tiling used to define a particular figure.

Net for the dodecahedron.

n-hedral tiling A tiling that has exactly n prototiles.

nonagon: A polygon having nine sides.

nonagon

noncollinear Not lying on the same line.

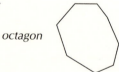

Points A, B, and C are noncollinear.

nonconsecutive Not in order or in sequence.
See consecutive for more information.

nonperiodic tiling A tiling which cannot replicate itself when translated in two or more non-parallel directions.

numerator In a fraction, the number written above the line.

O

obtuse angle An angle that measures between 90° and 180°.

obtuse angles

obtuse triangle A triangle having one obtuse angle.

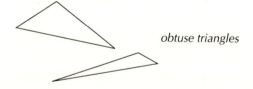

obtuse triangles

octagon A polygon having eight sides.

octagon

octant One of the eight spatial subdivisions created by the x, y, and z axes in 3-space.

odd number A whole number that is not divisible by two.

eg. 1, 3, 5, 7, 9 . . .
If n is any whole number, 2n+1 or 2n-1 will always designate an odd number.

opposite rays The two rays determined by a single point on a line that travel in different directions. Two rays are opposite rays if they are collinear and have only their endpoint in common.

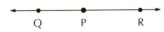

PQ and PR are opposite rays.

orbit The path that contains successive points in the iterations of a function of complex numbers.

order *In Fractal Geometry:* see generation.
In tiling: a numerical shorthand for a vertex net.

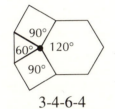

This vertex net is of order 3-4-6-4, 4-6-4-3. 6-4-3-4 or 4-3-4-6.

3-4-6-4

ordered pair A pair of numbers used to locate a point on the Cartesian coordinate plane. The first number gives the position relative to the x axis and the second gives the position relative to the y axis.

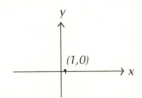

ordered triple A trio of numbers used to locate a point in 3-space. In order, they give positions relative to the x, y, and z axes, respectively.

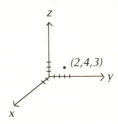

P

parallel *Lines* are parallel if they lie in the same plane and have no points in common.

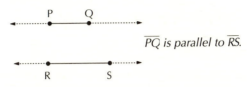

$\overleftrightarrow{AB}$ *is parallel to* $\overleftrightarrow{CD}$.

Segments are parallel if they lie on parallel lines.

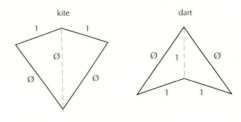

$\overline{PQ}$ *is parallel to* $\overline{RS}$.

parallelogram A quadrilateral in which opposite sides are parallel.

ABCD is a parallelogram.
See Appendix B for properties of a parallelogram.

pattern A repetition of units in an ordered sequence.

Penrose tiles The tiles discovered by Roger Penrose that can be used in pairs to form infinite nonperiodic tilings.

Penrose kites and darts

Penrose rhombuses

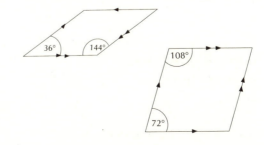

pentacle   See pentagram.

pentagon   A polygon having five sides.

pentagon

pentagram   The star shaped figure obtained by drawing all the diagonals of a regular pentagon.

Also called a pentalpha, pentacle, or pentangle.
The solid segments represent the pentagram. The dotted segments represent the corresponding regular pentagon.

pentalpha   See pentagram.

pentangle   See pentagram.

perimeter   The distance around a figure.

periodic tiling   A tiling that can duplicate itself when translated in two or more non-parallel directions.

perpendicular   *Lines* are perpendicular if they meet to form right angles.

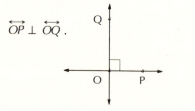

$\overleftrightarrow{OP} \perp \overleftrightarrow{OQ}$.

Segments are perpendicular if they lie on perpendicular lines.

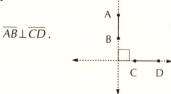

$\overline{AB} \perp \overline{CD}$.

perpendicular bisector   See bisect.

phi   The number naming the ratio that appears in the Divine Proportion (approximately 1.61803). The Greek letter phi (Ø) is used in this text, but the letter tau (τ) is also used by mathematicians to name this ratio.

Ø-Family Rectangles   The group of rectangles related to the Golden Rectangle or generated by Ø. They are: the square, the $\sqrt{Ø}$ Rectangle, the Golden Rectangle, the $\sqrt{4}$ Rectangle, the $\sqrt{5}$ Rectangle and the Ø+1 Rectangle.

plane   An undefined term in Geometry. A plane is often thought of as "a slice of space" having infinite length and width but no thickness.

plane group   A pattern resulting from repetition of a unit in two non-parallel directions to fill the plane. The unit is developed by manipulating a motif using one or more symmetry operations.
There are 17 plane groups.

point   An undefined term in Geometry. A point is often thought of as a location in space having no dimensions.

point group   A pattern that results from rotation of a unit about a pole.

polar grid   See grid.

pole   On a polar graph, the point corresponding to the origin.

polygon   A simple closed curve composed of line segments each intersecting exactly two others, one at each endpoint.

The line segments are called the sides of the polygon and the points of intersection are called the vertices of the polygon. A polygon is named by its vertices read consecutively in either a clockwise or counterclockwise direction.

This polygon could be named ABCDE. It could also be named DCBAE, among others.

primary colors In pigments, yellow, red and blue.

These are the essential hues from which all others are derived.

prime group Any one of the point groups developed from rotation and reflection.

The prime groups are named with a P followed by a numeral which indicates the parent base group, followed by the prime symbol.

A point group of order P2'

proportion A pair of ratios set equal to each other.

$$\frac{a}{b} = \frac{c}{d}$$

*In this proportion, **a** and **d** are called the extremes, and **b** and **c** are called the means of the proportion. These words refer only to the position of the terms.*

protractor A tool used for measuring the number of degrees in an angle.

protractor

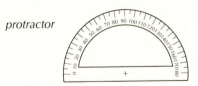

Pythagorean Theorem

$a^2 + b^2 = c^2$, where a and b name the lengths of the legs of a right triangle and c names the length of the hypotenuse.

Pythagorean triple or triad Three integers that have the property that the sum of the squares of two of them equals the square of the third.

eg. 3, 4, 5 is a Pythagorean triple since $3^2 + 4^2 = 5^2$ or $9 + 16 = 25$.

Q

quadratic equation An equation of the form $ax^2 + bx + c = 0$, where a, b and c are real numbers and $a \neq 0$.

quadratic formula

$$x = \frac{-b \pm \sqrt{b^2 - 4ac}}{2a}$$

where a, b and c are the coefficients of a quadratic equation $ax^2 + bx + c = 0$.

quadrilateral A four sided polygon.

quadrilaterals

R

radicand The quantity under the radical sign.

eg. in $\sqrt{15}$, 15 is the radicand.
In $\sqrt{3xy}$, 3xy is the radicand.

radius *Of a circle:* a line segment that joins the center of a circle to any point on the circle is *a* radius of the circle. The measure of any of these segments (all are congruent) is *the* radius of the circle.
See diagram next page.

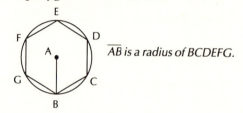

$\overline{OP}$ *is a radius of circle O.*

Of a regular polygon: If a regular polygon is inscribed in a circle, *the radius of the polygon is the radius of the circumscribed circle. A radius is a line segment from the "center" of the polygon to a vertex.*

$\overline{AB}$ *is a radius of BCDEFG.*

ratio   A comparison of numbers to each other.

The ratio of 2 to 3 may be expressed as 2:3 or simply 2 to 3 or as a fraction 2/3.

rational number   Any number that can be put into the form p/q where p and q are integers and q≠0.

ray   The figure formed by a point on a line and all the points on the same side of that point.

The point is called the endpoint of the ray, and the ray is named by its endpoint and any other point on the ray.

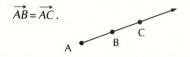

$\overrightarrow{AB} = \overrightarrow{AC}$.

reciprocal   *Of a rectangle:* a similar rectangle cut from the parent rectangle such that its length is the width of the parent rectangle.

ABCD ~ BCEF **a n d** *BCEF is the reciprocal of ABCD.*

Of a number: if *n* is any number then 1/n is its reciprocal.

Note: the product of any number and its reciprocal is always one. (n)(1/n)=1.

rectangle   A parallelogram whose angles are right angles.

ABCD is a rectangle.

Rectangle of Price   See √Ø Rectangle.

rectangular grid   See grid.

reflection   The symmetry operation that flips a motif, changing both its position and its handedness.

regular   *Polygon:* a polygon having all sides and angles congruent.

Tiling: A monohedral tiling in which the prototile is either a square, an equilateral triangle or a regular hexagon.

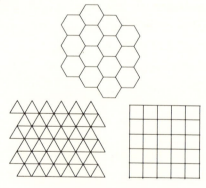

the regular tilings

rhombus   A parallelogram in which all sides are congruent.

ABCD is a rhombus.

right angle An angle that measures 90°.

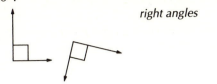

right angles

right triangle A triangle having one right angle.

right triangles

root See solution.

root rectangles The rectangles derived from the square such that the width measures one, and the length measures the square root of a natural number. These rectangles are the members of the family of Dynamic Rectangles.

rotation The symmetry operation that involves the turning of a motif about a point.

rotocenter A point about which a motif is rotated to create the point groups of symmetry.

S

scale invariance A term used to describe the property of self similarity in a given structure.

scalene triangle A triangle having no congruent sides.

scalene triangles

scaling The symmetry operation which combines translation with a proportional change in size of the motif.

secant A line that intersects a circle in exactly two points.

Line l is a secant with respect to circle P.

second One-sixtieth of a minute in angle measure.

$60'' = 1'$ or $3600'' = 1°$.

$18°32'21''$ reads 18 degrees, 32 minutes, 21 seconds.

sector A region in the interior of a circle bounded by two radii and the circle itself.

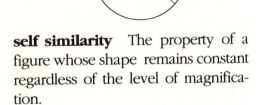

self similarity The property of a figure whose shape remains constant regardless of the level of magnification.

septagon A polygon having seven sides, sometimes called a heptagon.

septagon

sequence An ordered progression of terms.

A sequence may have a finite number of terms, eg. 1, 2, 3, 4, 5, 6; or an infinite number of terms, eg. 1, 1, 2, 3, 5, 8, 13, . . .

If the notation $S_1, S_2, S_3 ..., S_n ...$ is used to denote a sequence, the subscript names the position of the term. eg. S_{10} would be the tenth term.

set A collection of objects.

side *Of an angle:* see angle.

Of a line: either of the half planes formed by a line in the plane.

The shaded portions represent the two half planes. A and B lie on the same side of line l. A and C lie on opposite sides of line l.

Of a polygon: see polygon.

similar Having the same shape.

Two polygons are similar if corresponding angles are congruent and corresponding sides are in proportion.

$\triangle ABC \sim \triangle DEF$

$$\frac{AB}{DE} = \frac{BC}{EF} = \frac{AC}{DF}$$

simple closed curve A plane curve that can be traced without lifting pen from paper in such a way that the starting point and ending point are the same and no point is traced more than once.

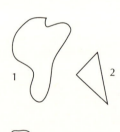

Figs. 1, 2 and 3 are simple closed curves. Figs. 4 and 5 are not.

solution Any quantity which may be substituted for a variable in an equation to produce a true statement.

eg. 3 is the solution for the equation x + 1 = 4.

square *Noun:* A rectangle in which all of the sides are congruent.

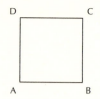

ABCD is a square.

Verb: To multiply a quantity by itself.

The number 2 used as an exponent indicates the operation of squaring. eg. $a^2 = (a)(a)$.

square root A number that when multiplied by itself produces a given number.

The operation of taking the positive square root of a number is indicated by a radical sign and the given number, under the sign, is called the radicand.
eg. $\sqrt{16} = 4$ since $4 \times 4 = 16$.

√Ø Rectangle (Square Root of Phi Rectangle): Any rectangle in which the ratio of the length to the width is √Ø to 1, or approximately 1.271.

straight angle An angle measuring 180°.

∠A is a straight angle, so called because its sides form a line.

Sublime Triangle See Golden Triangle.

summation sequence A sequence whose terms are arranged such that any term is the sum of the two preceding terms.

$u_1, u_2, u_3, ..., u_n, ...$ is a summation sequence if $u_i + u_{i+1} = u_{i+2}$ for all integers i.

supplementary angles Two angles the sum of whose measures is 180°.

The angles are said to be supplements of each other.

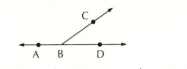

∠*ABC* and ∠*CBD* are supplementary.

T

tangent A line lying in the same plane as a circle that intersects the circle in exactly one point.

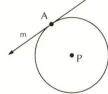

Line m is tangent to circle P at A. A is called the point of tangency.

tessellation A tiling.

tiling *Verb:* the process of covering the plane with closed units so that there are no gaps or overlaps.
Noun: the resulting pattern when the plane is tiled.

translation The symmetry operation which involves the sliding of a motif without changing its orientation.

trapezoid A quadrilateral with exactly one pair of parallel sides.

The parallel sides are called the bases of the trapezoid and the angles including either base are called the base angles. There are two pairs of base angles in a trapezoid.

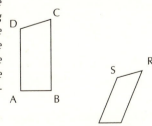

trapezoids

triangle A polygon having three sides.

Individual types of triangles are listed under their respective names.

Triangle of Price Any right triangle whose sides are in the ratio of 1; $\sqrt{\emptyset}$: $\emptyset$.

Also called the Egyptian Triangle.

Triangle of the Pentalpha See Golden Triangle.

triangulation The process of dividing an irregular shape into triangular shapes for purposes of finding area.

trihedral tiling A tiling that has exactly three prototiles.

truncate To cut off in a systematic fashion, especially to change one polygonal form to another.

U

uniform tiling A tiling in which all the vertices are congruent.

unit In symmetry, a pattern element determined by a motif used with one or more symmetry operations.

unit measure The length taken to represent the number one in a given set of circumstances.

V

vertex *Of an angle:* see angle.

Of a polygon: see polygon.

vertex net A diagram indicating the type and number of polygonal tiles surrounding a vertex in a tiling.

W

Whirling Square Rectangle Another name for the Golden Rectangle.

X

x axis The horizontal axis on the Cartesian coordinate plane.

Y

y axis The vertical axis on the Cartesian coordinate plane.

Index

A

B

C

trihedral, 141
vertex net(s) in, 137-138
Tondo, 83
Translation, in symmetry, 34
Triangle(s)
of Price, 140
special, in tiling, 140-142
Triangular tilings, 140-141
prototile in, 141

U

Universal Patterns, xii, 27,118, 140

V

Venn diagram, 115
Vertex nets, 137-138
Vesica Pisces, 115

X

X axis, 6

Y

Y axis, 6

DATE DUE

	1999		
AUG 2 4			
OCT 1 8 2002			

6-3-94

Demco. Inc. 38-293